Systems Engineering: Theory and Practice

Systems Engineering: Theory and Practice

Eden Brooks

STATES
ACADEMIC PRESS
www.statesacademicpress.com

Systems Engineering: Theory and Practice
Eden Brooks
ISBN: 978-1-63989-507-6 (Hardback)

STATES
ACADEMIC PRESS

Published by States Academic Press,
109 South 5th Street,
Brooklyn, NY 11249, USA

Cataloging-in-Publication Data

Systems engineering : theory and practice / Eden Brooks.
p. cm.
Includes bibliographical references and index.
ISBN 978-1-63989-507-6
1. Systems engineering. 2. System analysis. 3. Engineering. I. Brooks, Eden.
TA168 .S97 2022
620.001 171--dc23

Table of Contents

Preface

This book aims to help a broader range of students by exploring a wide variety of significant topics related to this discipline. It will help students in achieving a higher level of understanding of the subject and excel in their respective fields. This book would not have been possible without the unwavered support of my senior professors who took out the time to provide me feedback and help me with the process. I would also like to thank my family for their patience and support.

Systems engineering refers to the process of designing and managing complex systems over their life cycles. This discipline is an integration of engineering management and engineering. It is concerned with optimization methods, risk management and work processes of large and complex projects. It focuses on the formulation of various engineering concepts. This area of study has been divided into three types. These are product systems engineering, service systems engineering and enterprise systems engineering. The field finds its applications in other areas such as defense and aerospace, software development, electronics and communications, and information and technology. Configuration management, interface design, operations research, mechatronic engineering, proposal engineering, industrial engineering and cognitive systems engineering are some of the allied of this discipline. This book attempts to understand the multiple branches that fall under this discipline and how such concepts have practical applications. It also unfolds the innovative aspects of this field which will be crucial for the holistic understanding of the subject matter. Those with an interest in the systems engineering field would find this book helpful.

A brief overview of the book contents is provided below:

Chapter - What is Systems Engineering?

Systems engineering is the field of engineering that deals with the design, technical management, and complex operations of a system. It involves four specific types of engineered system - product system, service system, enterprise system and system of systems. This chapter has been carefully written to provide an easy understanding of systems engineering.

Chapter - Systems Architecture

A type of conceptual model that deals with the organization and management of structure and behavior of a system is termed as systems architecture. Architecture description language, lifecycle modeling language, etc. are a few concepts that fall under its domain. This is an introductory chapter which will briefly introduce about systems architecture.

Chapter - Systems Development

The process that deals with the design, testing and implementation of a new software program is called systems development. Its focus areas include axiomatic design, IPO model,

HIPO model, N2 diagram, etc. The topics elaborated in this chapter will help in gaining a better perspective of the related concepts of systems development.

Chapter - Systems Modeling Languages

Systems modeling language is referred to as a general purpose modeling language that assists the analysis, design and validation of a wide range of systems. Various diagrams are studied under it such as package diagram, use case diagram, activity diagram, requirements diagram, state diagram, etc. This chapter closely examines the subject of systems modeling languages to provide an easy understanding of the subject.

Chapter - Control Engineering

Control engineering is a discipline of engineering that is concerned with the application of automatic control theory to design systems in control environments. It includes control theory, stochastic control, open and closed loop system, feedback systems, etc. This chapter discusses the discipline of control engineering in detail.

Eden Brooks

1

What is Systems Engineering?

Systems engineering is the field of engineering that deals with the design, technical management, and complex operations of a system. It involves four specific types of engineered system – product system, service system, enterprise system and system of systems. This chapter has been carefully written to provide an easy understanding of systems engineering.

System

System can be considered as a physical entity which manipulates one or more input signals applied to it. For example a microphone is a system which converts the input acoustic (voice or sound) signal into an electric signal. A system is defined mathematically as a unique operator or transformation that maps an input signal in to an output signal. This is defined as $y(t) = T[x(t)]$ where $x(t)$ is input signal, $y(t)$ is output signal, T[] is transformation that characterizes the system behavior.

There are many classifications of systems based on parameter used to classify them. They are:

- Linear, non linear systems,
- Time variant, time invariant systems,
- Stable, unstable systems,
- Causal, non causal systems,
- Continuous time, discrete time systems,
- Invertible and noninvertible systems,
- Dynamic and static systems.

Linear and Nonlinear Systems

A linear system is one which satisfies the principle of superposition and homogeneity or scaling.

Consider a linear system 6 characterized by the transformation operator T[]. Let x_1, x_2 are the inputs applied to it and y_1, y_2 are the outputs. Then the following equations hold for a linear system:

$$y_1 = T[x_1], y2 = T[x_2]$$

Principle of homogeneity: $\mathbf{T\left[a*x_1\right] = a*y_1, T\left[b*x_2\right] = = b*y_2}$

Principle of superposition: $\mathbf{T\left[x_1\right] + T\left[x_2\right] = a*y_1 + b*y_2}$

Linearity: $\mathbf{T\left[a*x_1\right] + T\left[b*x_2\right] = a*y_1 + b*y_2}$

Where a, b are constants.

Linearity ensures regeneration of input frequencies at output. Nonlinearity leads to generation of new frequencies in the output different from input frequencies. Most of the control theory is devoted to explore linear systems.

Time Variant and Time Invariant Systems

A system is said to be time variant system if its response varies with time. If the system response to an input signal does not change with time such system is termed as time invariant system. The behavior and characteristics of time variant system are fixed over time.

In time invariant systems if input is delayed by time t_0 the output will also get delayed by t_0. Mathematically it is specified as follows:

$$\mathbf{y(t-t_o) = T\left[x(t-t_o)\right]}$$

For a discrete time invariant system the condition for time invariance can be formulated mathematically by replacing t as n*Ts is given as,

$$\mathbf{y(n-n_o) = T\left[x(n-n_o)\right]}$$

Where n_0 is the time delay. Time invariance minimizes the complexity involved in the analysis of systems. Most of the systems in practice are time invariant systems.

Note: In describing discrete time systems the sampling times $n*T_s$ are mentioned as n i.e. a discrete signal $x(n*T_s)$ is indicated for simplicity as x(n).

Stable and Unstable Systems

Most of the control system theory involves estimation of stability of systems. Stability is an important parameter which determines its applicability. Stability of a system is formulated in bounded input bounded output sense i.e. a system is stable if its response is bounded for a bounded input (bounded means finite).

An unstable system is one in which the output of the system is unbounded for a bounded input. The response of an unstable system diverges to infinity.

Causal and Non Causal Systems

The principle of causality states that the output of a system always succeeds input. A system for which the principle of causality holds is defined as causal system. If an input is applied to a system at time t=0 s then the output of a causal system is zero for t<0. If the output depends on present and past inputs then system is casual otherwise non casual.

A system in which output (response) precedes input is known as Non causal system. If an input is applied to a system at time t=0 s then the output of a non causal system is non zero for t<0. Such systems are referred as non-anticipative as the system output does not anticipate future values of input. Non causal systems do not exist in practice.

Continuous Time and Discrete Time Systems

A system which deals with continuous time signals is known as continuous time system. For such a system the outputs and inputs are continuous time signals. Discrete time system deals with discrete time signals. For such a system the outputs and inputs are discrete time signals.

Invertible and Non-invertible Systems

A system is said to be invertible if distinct inputs lead to distinct outputs. For such a system there exists an inverse transformation (inverse system) denoted by $T^{-1}[]$ which maps the outputs of original systems to the inputs applied. Accordingly we can write,

$$TT^{-1} = T^{-1}T = I$$

Where I = 1 one for single input and single output systems.

A non-invertible system is one in which distinct inputs leads to same outputs. For such a system an inverse system will not exist.

Dynamic and Static Systems

In static system the outputs at present instant depends only on present inputs. These systems are also called as memory less systems as the system output at given time is dependent only on the inputs at that same time.

Dynamic systems are those in which the output at present instant depends on past inputs and past outputs. These are also called as systems with memory as the system output needs to store information regarding the past inputs or outputs.

Properties of a System

A system is a physical device (or an algorithm) which performs required operation on a discrete time signal. A discrete time signal is represented as shown in figure below:

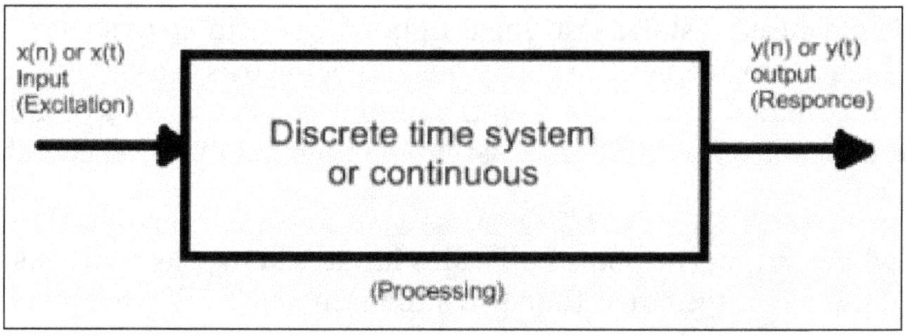

Discrete Time System

- Here x(n) input discrete time signal applied to the system. It is also called as excitation.
- The system operates on discrete time signal. This is called as procession of input signal, x(n).
- Output of the system is denoted by y(n). It is also called as response of the system.

Notation:

When input signal is passed through the system then it is represented by following notation:

$$x(n) \xrightarrow{T} y(n) \text{OR} \, y(n) = T[x(n)]$$

Here T is called as transformation operation. Similarly for continuous time system:

$$x(n) \xrightarrow{T} y(t) \text{ OR } y(t) = T[x(t)]$$

Example: A filter is good example of a system. A signal containing noise is applied to the input of the filter. This is an input signal to the system. The filter cancels or attenuates noise signal. This is the processing of the signal. A noise-free signal obtained at the output of the filter is called as response of the system.

System Engineering and its Processes

Systems engineering is a technique of using knowledge from various branches of engineering and science to introduce technological innovations into the planning and development stages of a system.

Systems engineering is not so much a branch of engineering as it is a technique for applying knowledge from other branches of engineering and disciplines of science in effective combination to solve a multifaceted engineering problem. It is related to operations research but differs from it in that it is more a planning and design function, frequently involving technical innovation. Probably the most important aspect of systems engineering is its application to the development of new technological possibilities with the specific objective of putting them to use as rapidly as economic and technical considerations permit. In this sense it may be seen as the midwife of technological development.

The word "systems" is frequently used also in other combinations, especially when elements of technological advance are not so important. Systems analysis is an example. Systems theory, or sometimes systems science, is frequently applied to the analysis of physical dynamic systems. An example would be a complex electrical network with one or more feedback loops, in which the effects of a process return to cause changes in the source of the process.

In general, a systems engineering approach is likely to differ from a conventional design approach by exhibiting increased generality in its basic logical framework and increased concern with the fundamental objectives to be achieved. Thus, at each stage the systems engineer is likely to ask both why and how, rather than merely how.

In addition to systems engineering, it is important to define systems themselves. The systems with which a systems engineer is concerned are first of all man-made. Second, they are large and complex; their component parts interact so extensively that a change in one part is likely to affect many others. Unless there is such interaction, there is little for the systems engineer to do, at least at the systems level; he can turn immediately to the components themselves. Another important characteristic of systems is that their inputs are normally stochastic; that is, the inputs are essentially random functions of time, although they may exhibit statistical regularities. Thus, one cannot expect to foresee exactly what the system will be exposed to in actual operation, and its performance must be evaluated as a statistical average of the responses to a range of possible inputs. A calculation based on a single precisely defined input function will not do.

Systems may also vary depending on the amount of human judgment that enters into their operation. There are, of course, systems such as electrical circuits, automated production equipment, or robots that may operate in a completely determinate fashion. At the other extreme, there are management and control systems, for both business and military purposes, in which machines in a sense do most of the work but with human supervision and decision making at critical points. Clearly these mixed human-machine systems offer the greatest variety both of possibilities and problems for the systems engineer.

Systems Engineering Techniques, Tools and Procedures

If a system is both large and complex in the sense in which these terms have been defined, it may be difficult to find out how it works. A large part of the content of systems engineering consists of techniques for the investigation of such relatively complex situations.

Modeling and Optimization

Perhaps the most fundamental technique is the flow diagram, or flowchart, a graphical display composed of boxes representing individual components or subsystems of the complete system, plus arrows from box to box to show how the subsystems interact. Though such a representation is very useful in an initial study, it is, of course, essentially qualitative. A more effective approach in the long run is construction of a so-called mathematical model, which consists of a set of equations, or sometimes simply of tables and curves, describing the interactions within the system in quantitative terms. It is not necessary for the mathematical model to be exact, as long as it serves its purpose. It frequently consists of piecewise linear approximations to basically nonlinear situations (i.e., a series of short straight lines that roughly approximate a curve). After the model has been constructed and checked, a number of mathematical techniques can be employed (including straightforward enumeration and computing) to find out what it says about the actual operation of the system. Often these calculations will have a probabilistic or statistical flavour.

When the components or subsystems interact significantly, it may be possible to achieve essentially the same final level of performance in many different ways. Limited performance by one subsystem may be offset by superior performance somewhere else. These optimization studies, called trade-offs, are important in suggesting how to achieve a given result in the most economical manner. They are equally valuable in suggesting whether or not the proposed result is in fact a reasonable goal to aim for. It may be found, for example, that a modest reduction in performance will permit radical savings in overall cost or, conversely, that the postulated equipment is capable of much better performance than is asked of it, at only nominally greater expense. (It may also turn out that the equipment can supply useful functions not originally contemplated. Computing systems, for example, can usually perform extra chores of record keeping at little increased cost). For all of these reasons, studies of such variables are an important part of systems engineering, both in the early exploratory phases of a project and in the final design.

Identifying Objectives

The formulation of suitable objectives for the final system is so important a part of the systems engineering process that it deserves special attention. It is, of course, always possible to state the general objectives of a system in vague or perfectionist terms. A sufficiently clear, precise, and comprehensive statement to serve as a basis for engineering studies, however, is another matter. Unless the situation has been well explored in the past, the real choices are not likely to be obvious when the work begins. Thus, the first task of the systems engineer is to develop as clear a formulation of objectives as possible. This usually involves computations and consultation with others interested in the system. Because the final statement must reflect value judgments as well as purely technical considerations, the systems engineer does not try to do this thinking alone but attempts to serve as a working focus and catalyst. Although issues of this sort naturally present themselves with particular force near the beginning of a systems study, they may recur in subsequent steps. The question of objectives is never really out of the systems engineer's mind.

The principal reason why a satisfactory statement of objectives may present such a problem is simply that most systems have multiple objectives, often in conflict with one another. In the design of transport aircraft, for example, there are a multitude of desirable characteristics, such as range, speed, payload, and safety, to be maximized, as well as undesirable characteristics, such as noise generation and air pollution, to be minimized. Because the same design cannot do the best job in all of these directions, a compromise achieving the most desirable overall performance is required. The most attractive compromise, which may require both study and ingenuity, is not likely to be found at all until some hard thinking has been done about what characteristics are really needed.

Especially difficult problems in defining objectives may arise when an existing technology is transplanted to some new disciplinary area. An example is the application of electronics such as computer techniques to medicine and education. It seldom happens in such cases that the best system is based on a simple one-for-one substitution, such as direct replacement of a classroom teacher by electronic hardware and computer-assisted instruction materials. It is much more likely that the most effective plan will turn out to be a rather complicated mixture of the old and the new. This conclusion, however, is likely to raise basic issues about the actual objectives of the new system, issues made no simpler by the interdisciplinary nature of the situation.

A Design Example

The design of the commercial transport plane mentioned above is an example of a systems engineering problem. In such a design the aerodynamic lift, the drag of fuselage and wings, the control apparatus, the propulsion system, and such auxiliary hardware as the landing gear all interact substantially. One element cannot be disturbed without affecting the others; all elements and aspects of the total system, and the interactions among them, must be considered. Thus, if designers make the fuselage fatter and the wings smaller in an effort to carry more payload at the same or higher speeds, a new control system might be needed because of the changes produced in the overall mechanical and aerodynamic characteristics of the vehicle. Stronger and heavier landing gear might be needed to withstand higher landing speeds. Almost surely, the new design would call for larger engines and fuel tanks to compensate for greater aerodynamic drag. Thus the designers would have lost ground in some respects and gained in others. The new plane might be more useful for short flights when not much fuel must be carried but less useful for long ones. Obviously, the system objective—the kind of airplane actually wanted—must control the direction of any such study.

The study becomes more interesting if a possible advance in basic technology is considered, such as an improvement in propulsion or aerodynamics, and it is desired to determine how it might best be applied in a new airplane design. The central systems engineering question then would probably encompass the relation between the available new plane characteristics and the needs of the existing air transportation system. Clearly, such an investigation can be made only by going to one of the upper levels in the systems hierarchy.

Finally, to operate the new airplane successfully, a whole series of supporting functions may be required, including routine checkout, maintenance, and spare parts supply, in addition to functions directly involved in the plane's flight. Though, under normal circumstances, these might readily be handled by the existing operating staff, it is part of the user orientation of the systems approach that the systems engineer is expected to anticipate any new requirements and make sure they are properly planned for.

To make adequate comparisons between competing objectives, a logical frame of reference, broad enough to include both, is needed. Thus, the systems engineer may study many situations in the framework of more than one system or a whole hierarchy of systems of steadily increasing generality. In the example of an airplane, the airplane itself is a possible system, as are the group of planes owned by one airline, the total number of airplanes in a particular country, and that nation's transportation facilities. Though the simplest system—the airplane itself—is a satisfactory reference for specific design problems, a more general framework may be needed to approach broader problems. Thus the individual airplane designer may seek to ameliorate air-traffic congestion by improving airplane takeoff and landing characteristics, permitting better utilization of existing airports. The airlines in turn may suggest construction of more and better airports. From the point of view of the transportation system as a whole, the best step might be to invest more money in high-speed rail facilities to carry part of the air-traffic load. In systems engineering the error of studying the problem within too narrow a framework is called the error of suboptimization.

User Orientation

The stress on systems objectives has one further consequence worth mentioning; i.e., that systems engineering is likely to be strongly user-oriented. This results naturally enough from the fact that systems objectives usually relate to overall performance, which is what the final user is interested in. The identity of technical interest between the systems engineer and the final user is usually marked; systems engineering is likely to give special consideration to such qualities as reliability, ease of maintenance, and convenience in operation. Moreover, the final step of a systems engineering project is typically an evaluation that attempts to find out how well the system works in the hands of the user.

Tools

The most obvious aspect of systems engineering tools is their great diversity. A leading text in the field states that "there is virtually no scientific discipline which may not be used in the design of some large-scale system" and singles out probability, mathematical statistics, computing, system logic, queuing theory, game theory, linear programming, cybernetics, group dynamics, simulation, information theory, servomechanism theory, and human engineering. To this list might also be added decision theory, nonlinear programming, some elements of econometrics, and communications theory as related to random processes.

In spite of this diversity, many of the tools of systems theory can be grouped under a few major headings. The analytic problems associated with optimization, for example, are a recurrent theme. Probability and statistics are also major areas that carry with them numerous more specialized topics, such as queuing theory and much of communications theory. Finally, computing is a major field for the systems engineer. If all else fails, direct calculation or simulation may produce the desired results.

These fields are all essentially mathematical in nature. The systems engineer may need knowledge and skills of other sorts as well. The mathematical model and associated objective functions that conventionally begin a systems analysis, for example, are only satisfactory to the extent that they adequately represent the real physical situation. The adequacy of a mathematical model in this sense, however, is a matter of physical or engineering rather than mathematical judgment. In some circumstances the systems engineer may also need to know something about experimental procedures in general and, in particular, about ways of maximizing the amount of information from a given testing program. This is particularly likely to happen in urgent high-risk projects, like space exploration or nuclear power generation, in which intermediate testing failures are bound to occur, and the systems engineer, as part of his overall responsibility, must decide what to do next. Even in simpler cases, in which the project should close in a final test and evaluation phase, the systems engineer is responsible for ensuring that this work is adequately carried out. A closely related question is the monitoring of testing procedures for routine quality control purposes. This also is logically part of the systems engineer's duties, which reflect a basic user orientation. When reliability is very important, as in space programs, it may be a major responsibility.

When the systems engineer's job is defined as including significant management responsibility, some acquaintance with modern management techniques is an obvious requirement. The techniques of particular importance are those that bear most directly on cost figuring and scheduling technological developments.

Finally, systems engineering is used in new situations that may involve the application of new discoveries in science or technology to existing technical areas or the application of known science or technology in new contexts. In either case the systems engineer obviously needs considerable substantive knowledge of the fields involved in order to make reliable plans.

It is apparent that no single person is likely to meet all of these specialized qualifications. Thus, systems engineering on any significant scale almost invariably involves a team approach.

Applications of Systems Engineering

Many useful systems are, in effect, modifications of previous designs. The proportions of the subsystems may be changed, but no substantial function has been added or left out. Chemical-processing plants and information systems, for example, are likely to be of this sort. The basic task of the systems engineer in such a situation is relatively straightforward; it is essentially a matter of reoptimizing the existing design to meet the new conditions.

In other circumstances, however, the basic systems concept represents a more radical break with the past. The new concept may involve the introduction of new functions or the realization of old functions in new ways. On the other hand, it may merely involve a radical change in system parameters (constants).

The Development of Radically New Systems

Radically new systems concepts are like inventions in ordinary engineering. Usually offering a substantial advance in overall performance, more than would be expected from a modest reproportioning of a known system, these clearly deserve special attention. On the other hand, in many cases it is impossible to predict accurately in advance of the development just what performance may be achievable in one or more of the critical elements of the new system. This leaves the systems engineer with a special problem in planning, which is usually addressed by establishing a minimum acceptable level of performance for the critical elements, with the rest of the system so arranged that whatever is realized beyond this level will appear as growth potential in the overall capabilities of the system. Thus definitive optimization studies may be postponed until the system is better understood.

The Nike Ajax missile system provides an example of the application of a radically new systems concept. The simple realization that the technology was available to provide a missile that could outmaneuver an enemy bomber taking evasive action was perhaps the systems invention in this case. (Guided missiles had been thought of before, but only for use against targets simpler than a rapidly maneuvering airplane). In a more limited sense, however, the key idea in the overall systems concept was probably the decision to use a command-guidance system, as opposed, for example, to a homing system.

In the command system, both the radars used for tracking the aircraft and the intercepting missile and the computer that calculates how the missile should change course are on the ground. Such a system requires a minimum of control apparatus in the missile. It also allows the missile to follow computer-determined paths that are aerodynamically favourable. This was an especially important consideration at the time in maximizing the range achievable with available propulsion systems. It also allows, through the computer, maximum flexibility in dealing with evasive action by the target. On the other hand, adequate accuracy from the ground-tracking system becomes increasingly difficult in a command system as the range is increased, whereas a homing system is not so limited. Thus the adoption of the command system implied a belief that the ground radars would be accurate enough to provide satisfactory interception even at the limits of the expected field of fire. As the development turned out, tracking accuracies were more than adequate for the purpose, and the surplus provided growth potential toward still longer ranges and higher probabilities of interception. (In other circumstances, of course, a different choice might have been better).

Long-term Systems Development

Thus far the description of systems engineering may seem to suggest that systems engineering efforts are essentially episodic. In many situations, however, there are important elements of continuity in what the systems engineer actually does. He is likely, in fact, to work on a series of similar problems as part of a long-term effort. In the case of telephone engineering, systems engineering groups have been set up formally as permanent parts of the overall organizational structure, each group having cognizance over some wide area of telephone technology. Thus continuity in this case may extend over many years.

Telephone engineering makes a particularly good example of this because of the importance of long-range forecasting for the telephone plant. One of the important responsibilities of the systems engineer in this field, for example, involves establishment of performance standards for new items in the plant. In most fields the systems engineer's role in this respect is comparatively modest because performance standards represent value judgments established by others. In telephony the systems aspects are more conspicuous because, at any given instant, the plant is a composite of new and old items, designed at different times in different ways to meet a variety of assumed operating conditions. The pieces must all work together. Any new apparatus or system must be compatible with what already exists. In addition, however, if an orderly evolution of the plant is to continue, the new apparatus must also be suited to the traffic and service needs of the future. Thus the systems engineer's responsibility, to provide a sort of long-term doctrine of performance standards, becomes very important.

Somewhat the same considerations affect the economics of new telephone systems. Here again a carefully developed forecast is essential. On the one hand, there are usually substantial economies of scale in communication apparatus. Thus it pays to take big steps in installing new equipment. On the other hand, capital charges on under-utilized equipment are likely to be excessive. Because the successive steps of design, production, and installation of a complicated system may take years, and because meanwhile both technology and service demands may change materially, the situation may increase in complexity, and the systems engineers may find that the choice of the most inviting system involves a calculated risk. They are responsible for judging the degree to which new possibilities should be exploited at any time. The principal result of assigning the telephone systems engineers a continuing organizational position and function is to put them into much more intimate contact with the technical frontier in communications. Their prime responsibility becomes that of monitoring the technical frontier to see what can be put to use in new operating systems.

Major technical advances may, of course, require many years between the original discovery or conception and the time when a practical design becomes feasible. The systems engineers keep in touch with the design and research work as it progresses throughout this interim period. They can exercise a valuable if indirect influence on the investigation simply by noticing weaknesses and errors that need correction for a project to be successful. It frequently happens that an extensive program of systematic measurements is called for before a new systems conception can be implemented, even when the basic conception is well established.

In many cases, technology suggests two or more competitive approaches to the same problem. If a continuing systems engineering organization exists, there is generally no need to make a premature choice between them. Rather, both lines can be followed until it is clear that one is superior or, perhaps, that each has its particular niche in the marketplace. In telephony, for example, this was the case in the long-continued rivalry between microwave transmission systems and those based on transmissions by a coaxial cable.

Applications to Government and Social Problems

Thus far, systems engineering has been dealt with in relation to two principal fields of application. One field is industry, in which the prospects of a further expansion of systems engineering appear bright. Existing applications furnish many good models, and it seems likely that such extensions can take place without raising many unusual problems. The other major field of application has been in military and space systems, and this may have been the principal force in shaping the systems engineering field. The possibility of new applications of systems ideas in nonmilitary areas of government also has come under consideration in the realm of worldwide basic social and economic problems. On the other hand, systems engineering as practiced in other contexts does not automatically transfer easily to this new environment. General interest in the subject dates, however, only from the late 1960s, and the field is incompletely explored.

In one experiment in the conversion of military systems engineering techniques, a U.S. state government contracted with four large aerospace companies (each of which had a substantial capability in systems engineering) to study the following four topics: (1) a statewide information-handling system, including a plan for implementation, (2) a program for the prevention and control of crime and delinquency, (3) a waste-management problem, and (4) a systems approach to basic transportation problems.

None of the four studies led to proposals that seemed attractive enough to be implemented by the state, and, in this respect, the experiment was a disappointment, although in view of the wide scope of the problems attacked and the limited effort called for by the study contracts, the result is not surprising. On the other hand, the experiment was useful in advertising the possibilities of systems analysis as applied to civil problems and in illuminating difficulties that may be encountered in making such applications. The experiment stimulated interest in the civil uses of systems methods both inside and outside the United States.

The potential applications of the systems approach to governmental activities are so numerous and so varied, in both the developed and developing worlds, that an exhaustive catalog would be out of the question. Nevertheless, it may be worthwhile to list a few of the most conspicuous possibilities. The most obvious class is made up of massive engineering attacks on very broad socioeconomic problems. These are the situations that seem to have most in common with the applications of systems methods in developing weapons. They include new transportation systems, comprehensive attacks on pollution, and radical reconstruction of urban areas. A concrete issue is the problem of power-plant location, an urgent question in many advanced and developing countries. The systems overtones are obvious. Generating stations are customarily interconnected so that a new plant has an impact on the availability of power over a considerable region, and, of course, the effects of thermal and atmospheric pollution from a given plant may also be widespread.

Other applications of systems analysis in the social sphere tend generally to be smaller and more easily treated. One class consists of the extension of military budgeting and methods of financial control to the nonmilitary world. Another application has been the use of systems analysis to support the technical aspect of foreign-aid programs. Other fields include the possible application of specific items of new technology in such areas as crime detection, fire fighting, and traffic control. Still other studies involve specific aspects of such subjects as housing and other types of building construction. Such studies attempt to be useful rather than broad or necessarily definitive for all time to come.

The applications of systems analysis in civil government obviously still have far to go before their potentialities are exhausted. On the other hand, there are many reasons why these potentialities can be realized only slowly, if at all. Some of them are related to the inherent difficulty of the problems presented—the wide range of both technical and social considerations that may enter certain decisions, for example. Others reflect some of the common characteristics of governmental structure, the necessary bureaucratization of functions, for example, or the frequent problem of overlapping jurisdiction. Still other problems reflect the fact that existing systems analysts are trained preponderantly in the physical sciences and engineering and thus may not be well matched to the socioeconomic issues they are likely to confront, though most systems analysis groups working in socioeconomic questions try to balance their strength by adding appropriate missing skills. The most common problem, however, is probably simply the need to build up an adequate basis for mutual cooperation between systems analysts and government.

As such an evolution proceeds, there may be an increasing tendency for individual systems analysts to become identified with the substantive area in which they work and to lose their special relations to systems analysis as a distinctive field. Thus, it may be the ultimate fate of systems analysis to disappear as a separate field and instead become an important constituent of the planning function required in many parts of modern society.

Systems Development Life Cycle

Life cycle is description of the distinct phases through which each product passes during its product life. This includes phases such as requirements definition, concept design, production, operation, maintenance, etc.

The period of time from when the idea for a product or service is conceived until the time it is no longer available for use. This includes phases such as a requirements phase, a development phase, an installation phase, an operation phase, and a retirement phase.

A life cycle is a plan, composed of several phases, aimed at maximizing the efficient development of a quality, usable product. A life cycle is a set of procedures, some required and others optional, which serve as a template for generating an individual design process. The plan is not meant to be a strict step by step process, but rather a flexible process insuring that users, designers, and management are directly involved in the development of the final product. In general, a life cycle requires the project team to plan the development process, obtain training, generate the deliverables, and obtain timely product team manager approval for continued development. The typical life cycle is composed of five phases: Investigation, User Requirements, Analysis, Design, and Implementation & Release. Each phase is defined by activities, deliverables, and checkpoints.

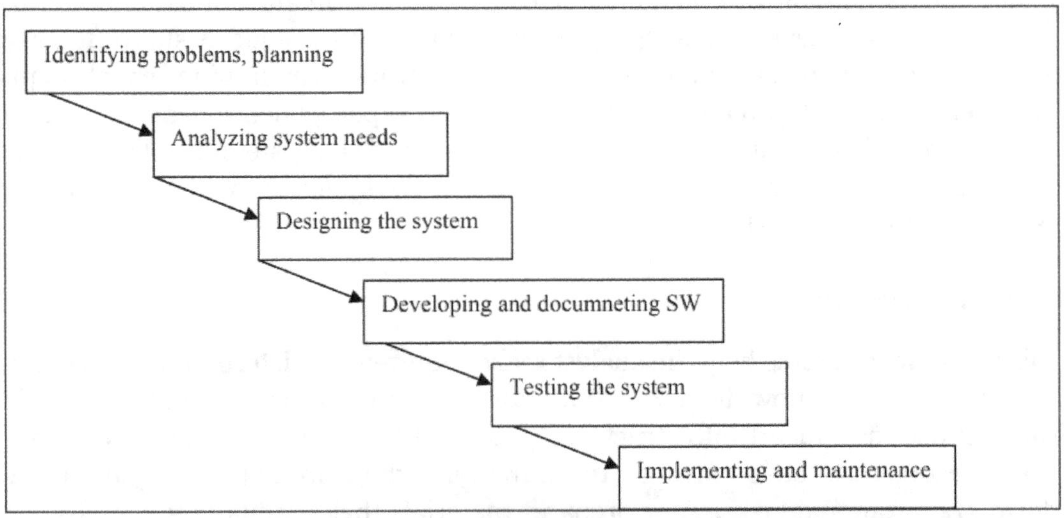

SDLC.

A little different meaning has term System development life cycle. It is usually seen as an organized, structured, methodology for developing, implementing, and installing a new or revised Computer Information System (CIS). Standard phases include investigation, analysis and general design, detailed design and implementation, installation, and review. The SDLC is a phased approach to analysis and design that holds that systems are best developed through the use of a specific cycle of analyst and user activities. Specialists disagree on exactly how many phases there are in SDLC. Every system development organization or textbook author uses slightly different life cycle model, with anywhere from three to almost twenty phases.

So what SDLC is? It is the process we would follow to build an information system. Basically it is the process companies go through to develop new information system. System development methodologies provide a framework for successful development of IS. Many information systems consulting companies develop their own methodologies. They use these proprietary methodologies as a means of differentiating themselves from the competition.

Identifying Problems and Planning

This is the first and very important phase of system development life cycle. This phase is critical to the success of the project.

Specialist identifies the need for a new or enhanced system. Information needs of the organization are examined and projects to meet these needs are proactively identified. The organizations IS needs may result from requests to deal with problems in current procedures, from the desire to perform additional tasks, or from realization that information technology could be used to capitalize on an existing opportunity. These needs can then be prioritized and translated into a plan for the IS department, including a schedule for developing new major systems. In smaller organizations determination of which systems to develop may be affected by ad hoc user requests submitted as the need for new or enhanced systems arises, as well as from formalized information planning system. Two additional major activities are also performed during the planning phase, the formal, yet still preliminary, investigation of the system problem or opportunity at hand and the presentation of reasons why the system should or should be not developed by the organization.

Analyzing System Needs

In this phase special tools help the analyst make requirement determinations. One such tool is the use of data flow diagrams. With DFD are input, processes and output of the business functions convert into structured graphical form. During analysis the analyst studies the organization´s current procedures and the information systems used to perform organizational tasks. They prepares materials that summarizes what has been found, provides cost and benefits analyses of alternatives and makes recommendations on what should be done. Once the recommendation is accepted the analyst can begin to make plans to acquire any hardware and system software necessary to build or operate the system as proposed.

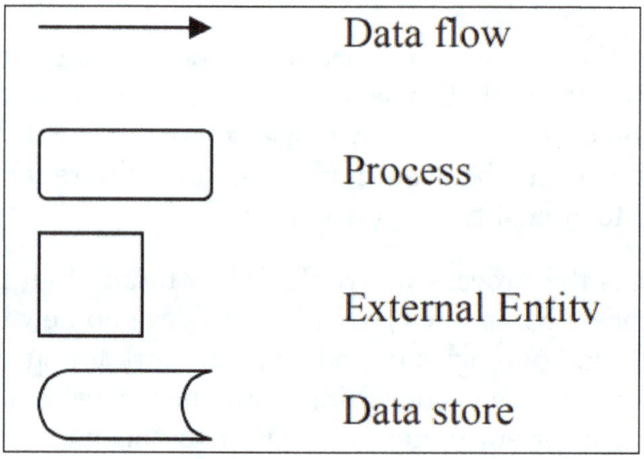

Data flow diagram symbols.

Designing the System

In this phase of SDLC analyst uses the information collected earlier to accomplish the logical design of the information system. The analyst designs accurate data entry procedures so that data going into the information system are correct. In addition the analyst provides for effective input to the information system by using techniques of good form and screen design.

Part of the logical design of the information system is devising the user interface. The interface connects the user with the system and is thus extremely important. Examples of user interface include keyboard or onscreen menus.

This phase also includes designing files or databases that will store much of the data needed by decision makers in the organization. A well-organized database is the basis for all information systems. In this phase analyst also works with users to design output that meets their information needs. They must design also controls and backup procedures to protect the system and the data and to produce program specification packets for programmers. Each packet should contain input and output layouts, file specifications and processing details – it may also include data flow diagrams, system flowcharts.

Developing and Documenting Software

In this phase the analyst works with programmers to develop any original software that is needed. Some of the structured techniques for designing and documenting software include structure charts or Nassi Scheiderman charts.

The system analyst uses one or more of these devices to communicate to the programmer what need to be programmed. During this phase the analyst also works with users to develop effective documentation for software including procedure manuals, online help, "Frequently asked questions", Read me files. Documentation tells users how to use software and what to do if software problems occur. Programmers have a key role in this phase because they design, code and remove syntactical errors from computer programs. If the program is to run in a mainframe environment job control language must be created.

Testing the System

So the system is build but is not ready for implementation. Before a company can run the system, it must be tested. It is much less costly to catch problems before the system is signed over the users. Some of the testing is completed by programmers alone, some of it by system analyst in conjunction with programmers. A series of tests to pinpoint problems is run first with sample data and eventually with actual data from the current system.

Implementing and Maintenance

This is the last phase of system development. The analyst helps to implement the information system. This phase involves training the users to handle the system. Some training is done by vendors, but oversight of training is the responsibility of the system analyst. In addition the analyst needs to plan for a smooth conversion from the old system to the new one. This process includes converting files from old formats to new ones or building a database, installing equipment and bringing the new system into production.

This phase of SDLC requires employees to use the new, unfamiliar system. This should be critical for final success.

A number of implementation approaches exists:

- Direct cutover implementation is when company quickly changes an old system for a new one. Usually a direct cutover implementation takes place over a week-end when employees are out of office. It can be very risky. It is risky in that, even after all of the testing, the new system may not function correctly.
- In parallel conversion approach a company installs a new system alongside its old system. This approach allows the organization and its employees to move to the new system gradually. The risk is that employees will continue to use old system.
- Pilot testing is next approach. In pilot testing a company installs a new system at only one location or in one department. Once the system has passed testing in this department is installed in all departments.

By some estimates the maintenance accounts much as 80 percent of the total costs of an information system. These estimates make sense when we consider that it might takes only a few months to a year to develop a new information system, but that system might be placed for decades. Maintenance includes, but is not limited to, correcting errors, backing up data or supporting end users. Some of the errors go undetected in the testing phase. It is when employees start using a new system that these errors come to the surface. Correcting these errors is one of the most important parts of system maintenance.

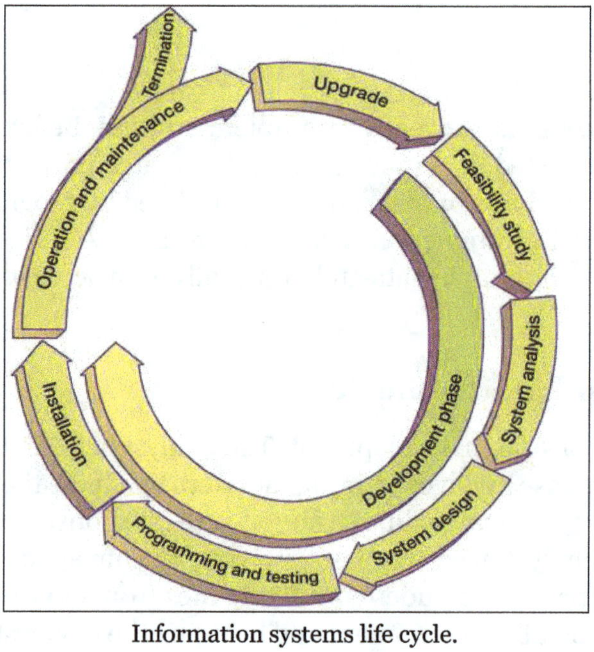

Information systems life cycle.

Model-based Systems Engineering

Model-based systems engineering (MBSE) is a methodology for systems engineering that uses visual modeling as the primary means of communicating information. Compared to a traditional document-based information exchange, MBSE eliminates the communication of unnecessary information by relying on abstract models that retain only the relevant data. This makes it easier for engineering teams to communicate design intent, understand the impact of design changes, and analyze a system's design before building it.

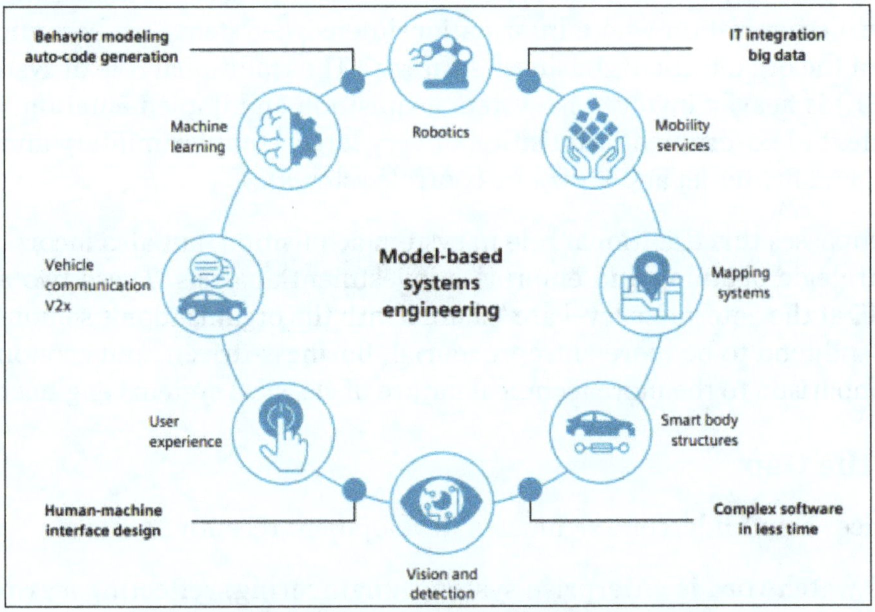

Model-based Systems Engineering eliminates the communication of unnecessary information by relying on abstract models that retain only the relevant data.

Need of Model-based Systems Engineering

In addition to improving communication among engineers and with project stakeholders throughout the System Development Life Cycle, MBSE helps with:

- Recording intellectual property related to system architectures, designs, and processes.
- Establishing scalable structures for troubleshooting.
- Managing the size and complexity of a project through visual models.
- Evaluating multiple system modifications simultaneously at minimal risk.
- Finding errors earlier in the system development life cycle.

Enterprise Systems Engineering

Enterprise systems engineering (ESE) is the application of systems engineering principles, concepts, and methods to the planning, design, improvement, and operation of an enterprise.

The purpose of traditional systems engineering (TSE) is to bring together a diversity of discipline experts to address a wide range of problems inherent in the development of a large, complex "single" system. Enterprise systems engineering (ESE) expands beyond this traditional basis to "consider the full range of SE services increasingly needed in a modern organization where information-intensive systems are becoming central elements of the organization's business strategy". The traditional role of systems engineering (SE) is heavily involved in system acquisition and implementation, especially in the context of government acquisition of very large, complex military and civil systems (e.g., F22 fighter jet and air traffic control systems).

ESE encompasses this traditional role in system acquisition, but also incorporates enterprise strategic planning and enterprise investment analysis. These two additional roles for SE at the enterprise level are "shared with the organization's senior line management, and tend to be more entrepreneurial, business-driven, and economic in nature in comparison to the more technical nature of classical systems engineering".

Closing the Gap

ESE practices have undergone significant development recently.

Today the watchword is enterprise systems engineering, reflecting a growing recognition that an 'enterprise' may comprise many organizations from different parts of government, from the private and public sectors, and, in some cases, from other nations.

Rebovich says there are "new and emerging modes of thought that are increasingly being recognized as essential to successful systems engineering in enterprises." For example, in addition to the TSE process areas, MITRE has included the following process areas in their ESE process to close the gap between ESE and PSE:

- Strategic technical planning.
- Enterprise architecture.
- Capabilities-based planning analysis.
- Technology planning.
- Enterprise analysis and assessment.

These ESE processes are shown in the context of the entire enterprise in the figure below. The ESE processes are shown in the middle with business processes on the left and TSE processes on the right. The TSE processes are well documented in many sources, especially in the ISO/IEC/IEEE 15288 standard.

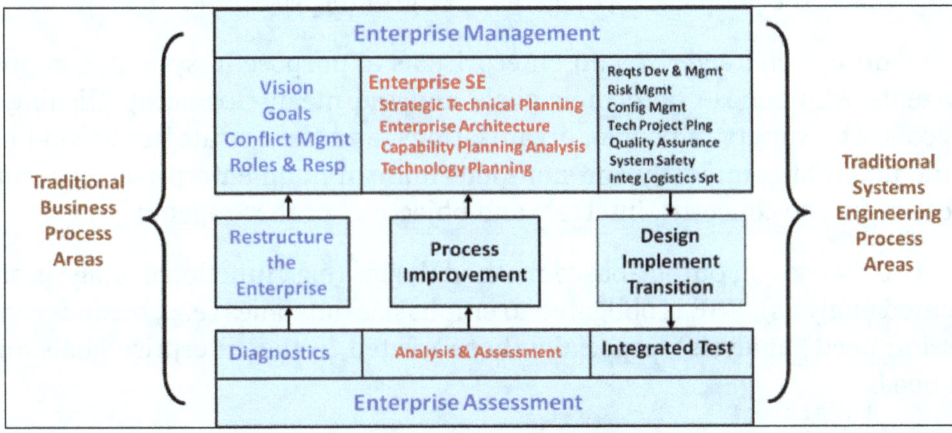

Enterprise SE Process areas.

SE is viewed by many organizations and depicted in many process definitions as bounded by the beginning and end of a system development project. In MITRE, this restricted definition was referred to as TSE. Many have taken a wider view seeking to apply SE to the "whole system" and "whole life cycle." For example, Hitchins sets out a holistic, whole-life, wider system view of SE centered on operational purpose. Elliott and Deasley discuss the differences between development phase SE and in-service SE.

In contrast to TSE, the ESE discipline is more like a "regimen" that is responsible for identifying "outcome spaces," shaping the development environment, coupling development to operations, and rewarding results rather than perceived promises. ESE must continually characterize the operational environmental and the results of enterprise or SoS interventions to stimulate further actions within and among various systems in the enterprise portfolio. Outcome spaces are characterized by a set of desired capabilities that help meet enterprise objectives, as opposed to definitive "user requirements" based on near-term needs. Enterprise capabilities must be robust enough to handle unknown threats and situations in the future. A detailed description of previous MITRE views on ESE can be found in a work by Rebovich and White.

Role of Requirements in ESE

TSE typically translates user needs into system requirements that drive the design of the system elements. The system requirements must be "frozen" long enough for the system components to be designed, developed, tested, built, and delivered to the end users (which can sometimes take years, and in the case of very large, complicated systems like spacecraft and fighter jets, more than a decade).

ESE, on the other hand, must account for the fact that the enterprise must be driven not by requirements (that rarely can even be defined, let alone made stable), but instead by continually changing organizational visions, goals, governance priorities, evolving technologies, and user expectations. An enterprise consists of people, processes, and technology where the people act as "agents" of the enterprise:

> "Ackoff has characterized an enterprise as a 'purposeful system' composed of agents who choose both their goals and the means for accomplishing those goals. The variety of people, organizations, and their strategies is what creates the inherent complexity and non-determinism in an enterprise. ESE must account for the concerns, interests and objectives of these agents."

Whereas TSE focuses on output-based methodologies (e.g., functional analysis and object-oriented analysis), ESE is obligated to emphasize outcomes (e.g., business analysis and mission needs analysis), especially those related to the enterprise goals and key mission needs.

Enterprise Entities and Relationships

An enterprise "system" has different entities and relationships than you might find in a product/service system. These can be usefully grouped into two categories: asset items and conceptual items. An example of an asset is hardware and software. Examples of conceptual items are things like analysis, financial elements, markets, policies, process, and strategy.

> An "enterprise system" should not be confused with the enterprise "perceived as a system." An enterprise system is a product (or service) system used across the enterprise, such as payroll, financial accounting, or enterprise resource planning applications, and consolidated data center, data warehouse, and other such facilities and equipment used across one or more organizations.

Products and services are sometimes treated as "assets" as shown in the figure below. This categorization of enterprise items comes from the semantic model (i.e., metamodel) used in the Troux Architect modeling tool for characterization and analysis of an enterprise architecture. Other enterprise entities of interest are things like information, knowledge, skills, finances, policies, process, strategy, markets, and resources, but these are categorized as "concept" items (in this particular schema). Further details on how to use this metamodel's entities and relationships are provided by Reese.

Table: Asset domain and concept domain categories for enterprise entities.

Application and Software Domain	Analysis Domain
Data Domain Document Domain Infrastructure and Hardware Domain IT Product Domain IT Service Domain Location Domain Organization Domain Product and Service Domain Services Portfolio Management Domain	Financial Domain General Domain Information Domain IT Architecture Domain Knowledge and Skill Domain Market Domain Policy Domain Process Domain Resource Domain Strategy Domain Timeline Domain Transition Domain

The application/software and infrastructure/hardware domains are likely the most familiar to systems engineers (as illustrated in the figure below). The application/ software domain contains things like the deployed software itself, plus applications, modules, servers, patches, functions, and messages. The infrastructure/hardware domain contains things like the hardware itself, plus networks and different kinds of hardware like computing hardware, cabinets, and network devices. There might be different subtypes of computing hardware like computers, servers, desktops, laptops, and mainframes. You can see from this elaboration of these domains that an enterprise architecture "schema" can be quite extensive in the kinds of things it can model.

The less technical domains would be things like policy, market, strategy, transition, financial, knowledge and skill, and analysis. In a typical enterprise architecture schema like this, there could be over a hundred types of modeling objects grouped into these domains. The examples give above are from the Troux Semantics metamodel used in the Troux Architect modeling tool for enterprise architecture activities. Other enterprise modeling tools have similar metamodels (sometimes called "schemas").

Enterprise Architecture Frameworks and Methodologies

Enterprise architecture frameworks are collections of standardized viewpoints, views, and models that can be used when developing architectural descriptions of the enterprise. These architecture descriptions can be informal, based on simple graphics and tables, or formal, based on more rigorous modeling tools and methods. ISO/IEC 42010 specifies how to create architecture descriptions.

These frameworks relate to descriptive models of an enterprise, with conventions agreed in particular communities. There are various frameworks and methodologies available that assist in the development of an enterprise architecture.

Functional Specification

A Functional Requirement (FR) is a description of the service that the software must offer. It describes a software system or its component. A function is nothing but inputs to the software system, its behavior, and outputs. It can be a calculation, data manipulation, business process, user interaction, or any other specific functionality which defines what function a system is likely to perform. Functional Requirements are also called Functional Specification.

In software engineering and systems engineering, a Functional Requirement can range from the high-level abstract statement of the sender's necessity to detailed mathematical functional requirement specifications. Functional software requirements help you to capture the intended behaviour of the system.

Example Functional Requirements

Functional Requirement No.	Function Requirement Description
FR 1	User should be able to enter sales Data
FR 2	Sales Reports should be generated every 24 hours
FR 3	API interface to Invoice System

Functional Requirements should include the following things:

- Details of operations conducted in every screen.
- Data handling logic should be entered into the system.
- It should have descriptions of system reports or other outputs.
- Complete information about the workflows performed by the system.
- It should clearly define who will be allowed to create/modify/delete the data in the system.
- How the system will fulfil applicable regulatory and compliance needs should be captured in the functional document.

Benefits of Functional Requirement

Here, are the pros/advantages of creating a typical functional requirement document:

- Helps you to check whether the application is providing all the functionalities that were mentioned in the functional requirement of that application.
- A functional requirement document helps you to define the functionality of a system or one of its subsystems.
- Functional requirements along with requirement analysis help identify missing requirements. They help clearly define the expected system service and behavior.
- Errors caught in the Functional requirement gathering stage are the cheapest to fix.
- Support user goals, tasks, or activities.

Types of Functional Requirements

The most common functional requirement types are:

- Transaction Handling.
- Business Rules.

- Certification Requirements.
- Reporting Requirements.
- Administrative functions.
- Authorization levels.
- Audit Tracking.
- External Interfaces.
- Historical Data management.
- Legal and Regulatory Requirements.

Example of Functional Requirements

- The software automatically validates customers against the ABC Contact Management System.
- The Sales system should allow users to record customers sales.
- The background color for all windows in the application will be blue and have a hexadecimal RGB color value of 0x0000FF.
- Only Managerial level employees have the right to view revenue data.
- The software system should be integrated with banking API.
- The software system should pass Section 508 accessibility requirement.

Non Functional vs. Functional Requirements

Here, are key differences between Functional and Nonfunctional Requirements:

Parameters	Functional Requirement	Non-Functional Requirement
What it is	Verb	Attributes
Requirement	It is mandatory.	It is non-mandatory.
Capturing type	It is captured in use case.	It is captured as a quality attribute.
End result	Product feature.	Product properties.
Capturing	Easy to capture.	Hard to capture.
Objective	Helps you verify the functionality of the software.	Helps you to verify the performance of the software.
Area of focus	Focus on user requirement	Concentrates on the user's expectation.
Documentation	Describe what the product does	Describes how the product works.
Type of Testing	Functional Testing like System, Integration, End to End, API testing, etc.	Non-Functional Testing like Performance, Stress, Usability, Security testing, etc.

Test Execution	Test Execution is done before non-functional testing.	After the functional testing.
Product Info	Product Features.	Product Properties.

Best Practice of Functional Requirement

Important best practice for developing functional requirement document is as follows:

- Do not combine two requirements into one. Keep the requirements granular.
- You should make each requirement as complete and accurate as possible.
- The document should draft all the technical requirements.
- Map all requirements to the objectives and principles which contributes to successful software delivery.
- Elicit requirements using interviews, workshops and casual communications.
- If there is any known, verified constraint which materially affects a requirement then it is a critical state that should be documented.
- It is necessary that you document all the assumption in the document.

Mistakes while Creating a Functional Requirement

Here, are some common mistakes made while creating function requirement document:

- Putting in unjustified extra information that may confuse developers.
- Not putting sufficient detail in the requirement document.
- You add rules or examples, scoping statements or objectives anything except the requirement itself.
- Left out a piece of important information that is an absolute must to fully, accurately, and definitively state the requirement.
- Some professionals start to defend the requirements they have documented when the requirement is modified, instead of finding the correct truth.
- Requirements which are not mapped to an objective or principle.

Functional Analysis in Systems Engineering

Functional Analysis is a fundamental tool of the design process to explore new concepts and define their architectures. When systems engineers design new products, they perform Functional Analysis to refine the new product's functional requirements, to map its functions to physical components, to guarantee that all necessary components are listed and that no unnecessary components are requested and to understand the relationships between the new product's components. The chapter begins with the definition of the role of Functional Analysis in conceptual design and then proceeds with the description of a proposed methodology and with the presentation of its applications at subsystem, system and system of systems levels. Eventually some conclusions are drawn.

The design process, in particular the design process of complex systems, can be split into three phases:

- Conceptual design.
- Preliminary design.
- Detail design.

Even though Functional Analysis applies to every phase of the design process, it turns out to be particularly useful during conceptual design, when there is still a wide range of potentially feasible solutions for the future product. The precious role of Functional Analysis consists in individuating as many available options as possible, without forgetting any ideas that may offer significant advantages. In the remainder of the chapter we refer specifically to the application of Functional Analysis during the conceptual design phase to explore complex systems.

Functional Analysis in Conceptual Design

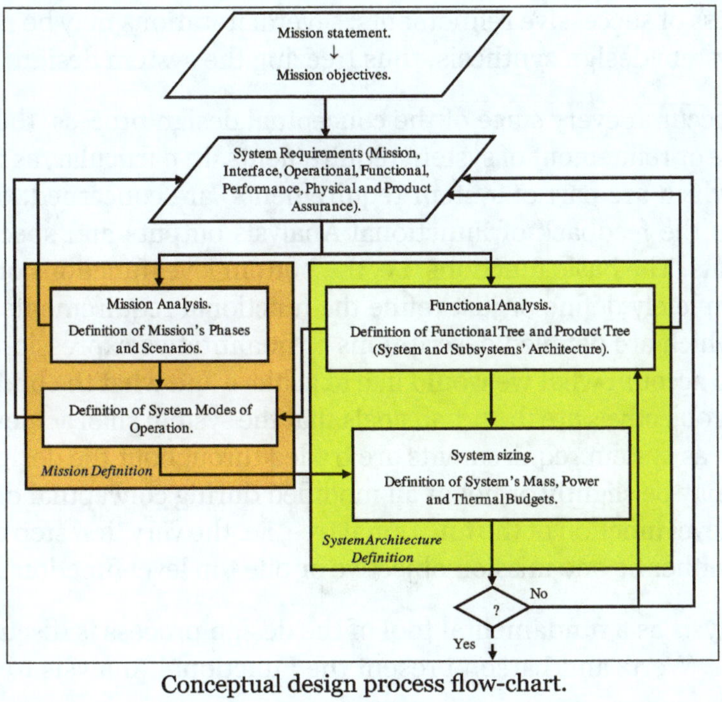

Conceptual design process flow-chart.

The conceptual design process is schematically illustrated in figure, where the role of Functional Analysis is highlighted, as well as its interactions with all other building blocks of the conceptual design methodology. Starting from the mission statement, the mission objectives can be derived. Once the broad goals of the system, represented by the mission objectives, have been established, the system requirements can be defined. On the basis of the system requirements, the conceptual design process evolves through the system architecture and the mission definition. The system architecture definition consists of the two main tasks: Functional Analysis and System Sizing.

Primary results of Functional Analysis are the functional tree and the product tree: the former identifies the basic functions, which the system has to be able to perform, while the latter individuates all system physical components, which are able to carry out the basic functions. In other words, these components may be the equipments or the subsystems, which constitute the whole system. Once the components of the product tree have been identified, it is possible to investigate how they are connected to form the system. It is thus possible to develop both the functional block diagram (secondary or additional result of Functional Analysis) and the physical block diagram of each subsystem and of the whole system. In order to complete the system architecture, the definition of the system budgets (mass, electric power, thermal power budgets, etc.) has to be carried out. However this task can be fulfilled only after the system modes of operation have been established. The modes of operation are part of the mission definition and can in their turn been set up only after the subsystems and their equipments have been identified. Once both the mission and the system architecture have been preliminary defined, before proceeding any further with the system design synthesis, it is important to verify whether or not all system requirements have been satisfied. Being the design activity typically a process of successive refinements, several iterations may be necessary before achieving the system design synthesis, thus freezing the system design.

Iterations may occur at every stage of the conceptual design process, thus resulting in a continuous trade or refinement of system requirements. In particular, as far as functional requirements (which are part of system requirements) are concerned, their refinement is mainly due to the feedback of Functional Analysis outputs and specifically of functional tree outputs. The basic functions, i.e. the bottom level functions of the tree, are in fact used to completely define or just refine the functional requirements. Unlike system requirements, which are detailed descriptions or quantitative expressions of the system itself, taking into account what we would like to achieve and what the budget allows us to achieve, mission objectives are the broad goals that the system shall achieve to be productive. Thus, whereas system requirements are traded throughout the design process, mission objectives may be slightly or not at all modified during conceptual design. For these reasons the top level function of the functional tree, i.e. the very first step of the Functional Analysis, can either be one mission objective or one top level functional requirement.

Functional Analysis as a fundamental tool of the design process is discussed by a number of references. Wertz and Larson present the Functional Analysis to decompose the functional requirements and focus only on one single task of Functional Analysis, i.e. the functional tree. NASA and ESA consider Functional Analysis as the systematic process of identifying, describing, and relating the functions a system has to be able to perform, in order to be successful, but does not consider it as a design tool to address how functions will be performed, i.e. to map functions to physical components. Particular emphasis is given to the possibility of capturing the technical requirements by performing Functional Analysis. In contrast we present Functional Analysis both to define the system functional architecture, through the development first of the product tree and then of the functional block diagram, and to define or refine the functional requirements, through the accomplishment of the functional tree. The following section describes into the details the tasks of the proposed Functional Analysis methodology.

Methodology

Starting from the mission objectives/top level system requirements or directly from the mission statement, the Functional Analysis allows identifying the physical components, the so-called building blocks, which constitute the future product, and how they are interrelated to build up the functional architecture of the future product. Moreover through Functional Analysis the functional requirements can be defined or anyway refined.

In conceptual design Functional Analysis can be applied at different levels: subsystem level (like the avionic subsystem of an aircraft, consisting of various pieces of equipment), system (like a satellite consisting of various subsystems) and system of systems level (like a Moon base, consisting of various systems. According to the considered level, the physical components or building blocks, which make up the future product, are therefore equipments, subsystems or systems.

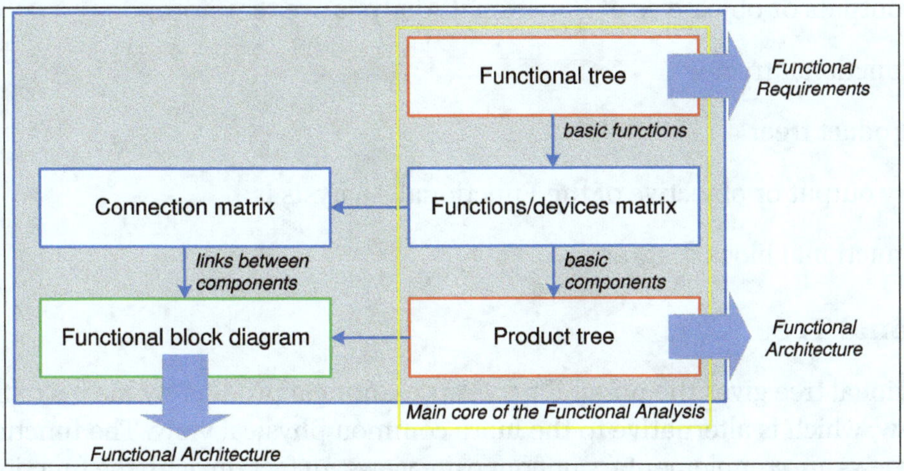

The Functional Analysis.

Figure shows the flow-chart of the proposed Functional Analysis methodology, illustrating all its tasks, how the various tasks are related one another and the inputs/outputs of each task.

- Functional tree.
- Functions/components (or functions/devices) matrix.
- Product (or physical) tree.
- Connection matrix.
- Functional block diagram.

On the basis of the mission objectives/top level system requirements the functional tree has to be developed as first step of Functional Analysis. Once the basic functions have been identified and the functional tree has therefore been completed, the functions/components matrix can be built and the basic components of the product tree can be individuated. Once the basic components have been determined, both the product tree and the connection matrix can be completed. Eventually, knowing all components (thanks to the product tree) and their relationships, the functional block diagram can be fulfilled.

As highlighted in figure, the main core of Functional Analysis is made up of the functional tree, the functions/devices matrix and the product tree. In fact through the functional tree and particularly through the identification of the basic functions, the functional requirements of the future product can be defined or refined, and through the product tree the building blocks of the future product can be determined, thus laying the major groundwork for the definition of the functional architecture of the future product. The functional architecture can then be completed once the relationships between the various components are clearly identified, i.e. after developing the connection matrix and the functional block diagram.

Primary outputs or objectives of Functional Analysis are therefore:

- Functional tree.
- Product tree.

Secondary output or objective of the Functional Analysis is:

- Functional block diagram.

Functional Tree

The functional tree gives the possibility of representing a product by means of the functional view, which is alternative to the more common physical view. The functional and physical views are complementary not opposite views. In fact through the functional view we look at a new product asking ourselves "what does it do?", while through the physical view we look at a new product asking ourselves "what is it?", which is without any doubts the most immediate question that arises in our mind, when looking at something that is unknown. Both views are valid as they are fundamental approaches to analyze complex systems by subdividing them into parts, characterized by a poor or high level of details, depending on the need of thoroughness and/or on the level of the analysis itself.

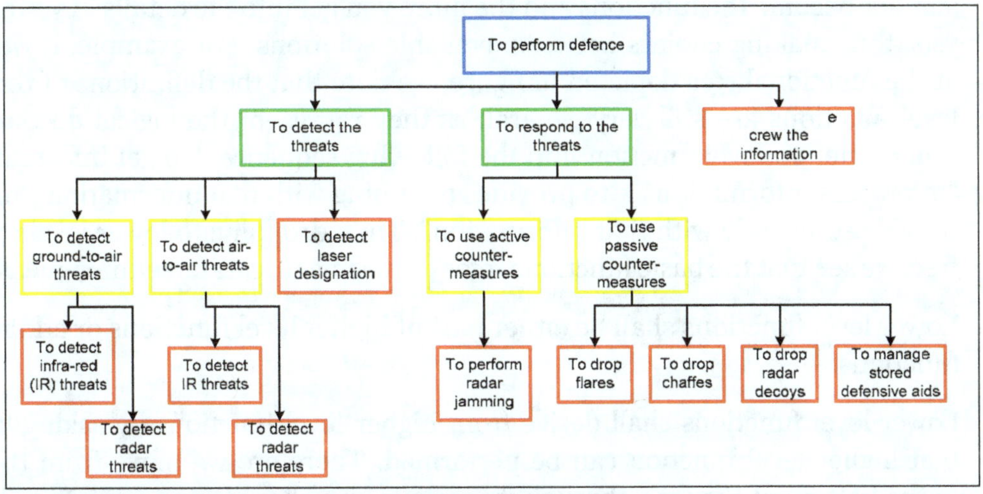

Example of functional tree.

The functional tree allows splitting the higher level functions, which stem from the mission objectives/top level system requirements, into lower level functions and eventually it allows identifying the basic functions that have to be performed by the future product. Higher level functions are complex functions that have to be decomposed into simpler functions, i.e. lower level functions, in order to accomplish the analysis. Therefore, starting from the socalled top level function, the functional tree generates various branches, which move from the most complex function to the basic functions, i.e. those functions at the bottom of the tree that cannot be split any further. Main output of the functional tree is therefore the identification of the basic functions through the decomposition of the higher level functions. The basic functions help defining or refining the functional requirements of the future product, as each basic function can be rewritten as a functional requirement. As an example, the basic function of figure "To detect infra-red (IR) threads" can be rewritten as "The system shall be able to detect infra-red (IR) threads". Figure shows an example of functional tree. The top level function is "To perform defence", particularly the defence of a military aircraft. The blue box represents the top level function, while the red boxes represent the basic functions. Starting from the top level function and getting down to the basic functions, two successive levels of functions decomposition can be noted: the first and the second level functions.

In order to carry out the functional tree, the next rules have to be followed:

- Each function shall be expressed by means of verb and noun.
- The definition of each function shall be as general as possible. Pursuing maximum generality, when describing functions, allows fostering the search of alternative solutions, in order not to forget any valuable options. This fundamental rule can be satisfactorily applied at the highest levels of the functional tree.

However the lower are the levels of the tree, the less general are the functions' definitions. It is true in fact that the more you go down the three branches, the simpler become the functions and the more you get into the details of your analysis, thus making choices between available solutions. For example, if we look at the functional tree depicted in figure, we note that the definitions of the first level functions are still very general, as they represent the logical decomposition of the top level function into the following sequence: "to get information", "to process information" , "to provide something with that information" and "to provide somebody with that information". Then dropping to lower levels of the tree, we see that the basic functions refer to specific threats or counter-measures.

- Lower level functions shall be either part of higher level functions or additional functions.
- Lower level functions shall derive from higher level functions by asking "how" that higher level function can be performed. Therefore we move from the top to the bottom of the tree, through its various branches, asking ourselves "how". Viceversa we move from the bottom to the top of the tree by asking ourselves "why". Looking again at the example reported in figure, we may decompose the top level function "to perform defence" by asking ourselves: "how can the defence (of a military aircraft) be performed?". The answer to this question, that will thus represent the first subdivision of the top level function, may be: "to detect the threats" (i.e. "to get information"), "to respond to the threats" (i.e. "to process information" and "to provide something with that information") and "to display to the crew the information" (i.e. "to provide somebody with that information").
- In case functions cannot be decomposed any further, they shall be reported at the bottom of the tree as basic functions. As an example, in figure the three functions, that are located in the row immediately beneath the top level function, represent the first decomposition of the top level function itself. Two out of these three functions are first level functions and are further subdivided, while the remaining function is already a basic function, as it cannot be split into other sub-functions. It is worth noting that the choice of carrying on decomposing a certain function or of stopping decomposing it depends on the level of details of the whole analysis.
- The individuation of basic functions shall depend on the level of details of the whole analysis. This implies that, if the target of Functional Analysis is, for instance, the definition of the functional architecture of a system at main equipments level, the basic functions of the functional tree shall be those functions related to specific equipments, like the example shown in figure.
- If we conceive different (physical) solutions at a certain level of the functional tree, the tree shall change from that point downwards but not upwards. This implies that, starting from the same mission objective, different functional trees can be generated not only because different people are working at it but because

at a certain level of the tree (typically at lower levels) you may be obliged to make choices between alternative solutions. In this case, depending on the number of available options, a few functional trees shall be developed: they will be exactly the same from the top level function up to a certain level and will be different from that level to the bottom.

Eventually, getting back to the comparison between the function and physical views, the main advantages/criticalities of the functional view (i.e. typical approach of the functional tree) are reported hereafter.

The most significant advantages can be summarised as follows:

- The development of the functional tree, starting from mission objectives/top level system requirements, implies a thorough analysis of the mission objectives/top level system requirements themselves. This guarantees that the product, defined on the basis of the functional tree, meets all customer's needs and this is particularly important, if we remember that the functional tree is a design tool, useful to develop a new product. It is worth remembering here that, when we carry out the functional tree, we know very little about the new product. We just know the mission objectives and typically we have a preliminary draft of the system requirements but we ignore all elements that will constitute the new product. Thanks to the functions/devices matrix and then to the product tree, we will be able to say what elements will constitute the new product.
- The abstract view, typical of the functional tree, fosters the search of alternative solutions, thus avoiding biased choices.
- The functional view is absolutely coherent with the systems engineering view, which looks at the system as the integration of various elements.

The most significant criticalities can be summarised as follows:

- Starting from the same mission objective/top level system requirement, different functional trees can be developed, depending on the people working at it and on the envisaged solutions. It is clear therefore that carrying out a functional tree is a typical design activity, which requires the widespread knowledge of the systems engineering designer, whose mind is not confined to any specific discipline but can embrace the whole multi-disciplinary system as integration of various parts.
- As typically the available options may be many, the main risk resides in the possibility of forgetting some concepts that may offer significant advantages for the future product.

Functions/Devices Matrix and Product Tree

Once the basic functions have been identified, it is possible to choose the components that will perform those functions by means of the functions/components (or functions/devices) matrix. The functions components matrix is therefore used to map functions to physical components.

The functions/components matrix can be built simply by matching the bottom of the functional tree, consisting of all basic functions, with one column of components able to perform those functions. Starting from the column containing the first basic function under consideration, the component able to perform that function can be defined by simply answering the question: "which component is able to perform this function?". This component can then be written down in the first row of the column of devices. The same process applies to all basic functions. Starting from the analysis of the first basic function, new components progressively fill in the column of devices. Eventually all basic components are determined. Following the procedure reported above, we take into account the first basic function on the left hand side of the functions/devices matrix, "to detect infra-red (IR) threats". If we ask ourselves which component or better which equipment is able to perform this function, we may answer that both the missile warning receiver and the infra-red (IR) warning receiver are able to fulfil the task. Then we write down both equipments in two separate rows of the functions/devices matrix and tick the intersections between these rows and the column of the basic function under consideration. Applying the same procedure, we gradually complete the functions/devices matrix, thus identifying all basic equipments.

Remembering the logical decomposition of the top level function reported in figure ("to get information": "to detect the threats", "to process information": "to respond to the threats", "to provide something with that information": "to respond to the threats" and "to provide somebody with that information": "to display to the crew the information"), we note that:

- Sensors are equipments that detect threats.
- Processors are equipments that process information.
- Passive or active counter-measures are equipments that respond to threats.
- Displays are equipments that display to the crew the information.

Thanks to the functions/devices matrix we now know the basic components or building blocks, which constitute the future product. By simply grouping together the basic components, the product or physical tree of the new product can be generated. Unlike the functional tree, which has a typical top-down approach, the development of the product tree follows a straightforward bottom-up process. As we do know, according to the considered level, i.e. subsystem, system or system of systems level, the building blocks are respectively equipments, subsystems or systems. In case, for instance, the building blocks are equipments, they may be grouped into subsystems to form the whole system or, better, the product tree of the whole system, like illustrated in figure.

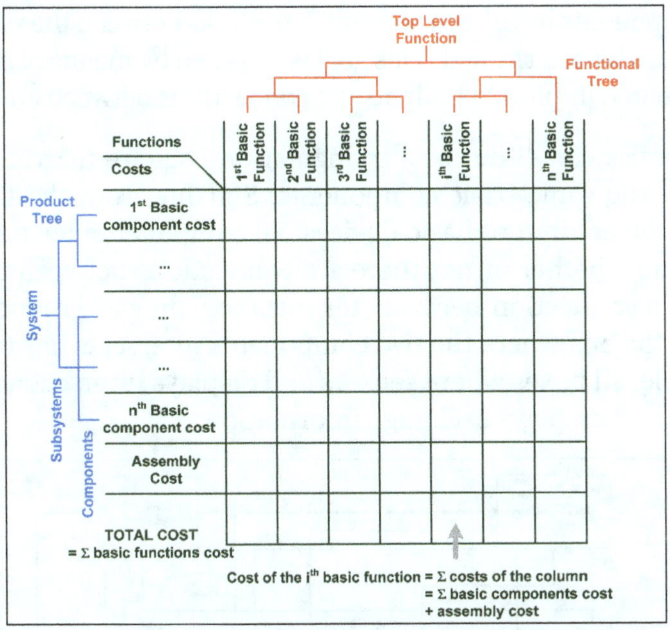

Product tree and costs/functions matrix.

In particular figure also shows the so-called functions/costs matrix, which is exactly the same as the functions/devices matrix except for the fact that here there is a column of costs instead of a column of devices. Quite obviously the functions/costs matrix can be built only after the functions/devices matrix, i.e. once the basic components have been identified. Main difference of the functions/costs matrix with respect to the functions/devices matrix lies in the consideration of the assembly cost. In fact, apart from the cost of each single basic component, the cost due to the assembly has to be taken into account, in order to estimate the cost of each single function and consequently the cost of the whole product.

Connection Matrix and Functional Block Diagram

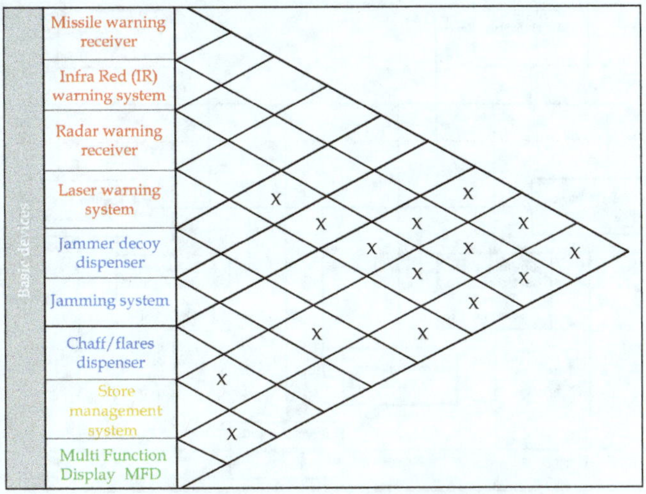

Example of connection matrix.

Once the basic components have been identified, the links between the various components within the system can be determined. This goal is achieved by means of the connection matrix, which, as the name implies, highlights the connections between all building blocks.

The connection matrix can either be a triangular or a square matrix, where both rows and columns have the same basic components. Starting from the first row and then proceeding down the column of basic devices, all components have to be analyzed, in order to understand whether or not there are connections between them. In case two components have a connection because, for instance, they are requested to exchange information, then the box where the two components intersects has to be ticked. As we can see, for example, all boxes where sensors and displays intersect have been ticked to show that sensors and displays exchange information.

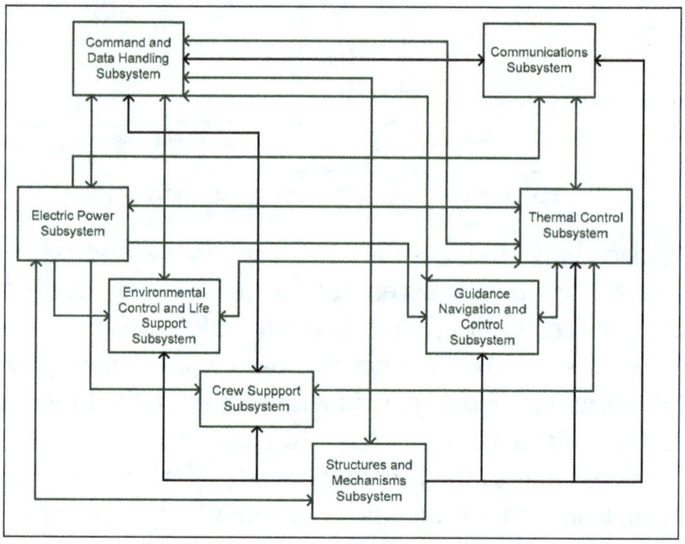

The functional block diagram.

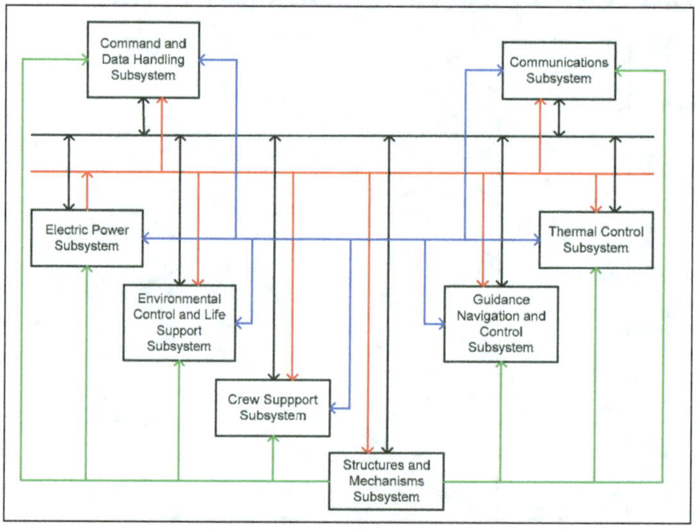

The physical block diagram.

It is worth underlying that nothing is said explicitly about the nature of the connections. The type of connection between all equipments is a pure signal of information between sensors, processors, passive or active counter-measures and displays.

A different representation of the same concept, expressed by the connection matrix, is obtained through the functional block diagram, where building blocks, that need to be connected, are linked through point-to-point connections. In case these connections are arrows and not just simple lines, the functional block diagram provides the reader with additional information, if compared to the connection matrix, as it highlights not merely connections but where these connections are pointing to, i.e. if they are half duplex or full duplex connections. Just by looking at a functional block diagram, it is therefore possible to understand that, for instance, sensors are transmitting information to displays and not vice versa. Like in the connection matrix, also in the functional block diagram nothing is said about the nature of the connections, which may be, for instance, power, signal or fluid lines. This information is instead provided by the physical block diagram, which may be considered as complementary to the functional block diagram.

figure shows an example of functional block diagram for a complex system, consisting of various subsystems. This system, named Permanent Habitable Module (PHM) is the first module of a permanent future human settlement on the Moon, designed to sustain the presence of three astronauts on the lunar surface. All main subsystems are highlighted in different boxes and the connections between them are shown. For sake of clarity, figure illustrates the physical block diagram of the same system presented in figure. Four different types of connections between the building blocks have been envisaged: structures, power, signal and fluid lines.

Structures guarantee, specifically by means of secondary and tertiary structures, the anchorage of all subsystems and particularly of all their equipments to the primary structure. A power line supplies the various building blocks with the necessary power. As far as the signal lines are concerned, it is worth noting that, unlike the functional block diagram where there are point-to-point connections, in the physical block diagram there is a main bus to transmit commands and receive feedbacks to/from the various subsystems. Eventually the building blocks that need an active cooling interface to dissipate heat are connected by a ducting fluid line with the Thermal Control Subsystem.

Applications

As the Functional Analysis can be applied at different levels, three different examples of applications of the methodology are presented in the following sub-sections:

- Functional Analysis at subsystem level to define the avionic subsystem of an aircraft.
- Functional Analysis at system level to define a satellite in Low Earth Orbit (LEO).

- Functional Analysis at system of systems level to define a permanent human Moon base.

Functional Analysis at Subsystem Level: The Avionic System of a very Light Business Jet Aircraft

The VLBJ avionics has been chosen as useful example because of its new functionalities and characteristics, which are not implemented in the avionic system of other civil aircraft. In fact VLBJs are designed to be certified as single pilot operations. This is made possible by advanced avionics automation, functional integration and easy-to-use capability.

Considering the aircraft mission profile, the environment where the aircraft will have to operate (air traffic control, landing and takeoff aids, navigation aids) and passengers and pilot requirements, the following macro-functions can be identified:

- To allow navigation.
- To perform flight controls.
- To allow communications.

For sake of simplicity only the macro-function "to allow navigation" will be dealt with here, in terms of functional tree and functions/devices matrix.

The complete functions decomposition of the top level function "to allow navigation" is reported hereafter.

To allow navigation:

- To acquire data:
    - To identify weather situation.
    - To detect magnetic field.
    - To acquire surrounding terrain altitude.
    - To acquire airplane data.
    - To acquire airport data.
    - To acquire flight plan data:
        - To acquire data about navigation aids (VOR-DME) ground station:
            - To memorize waypoints (VOR-DME stations).
            - To acquire radial and distance.
            - To calculate flight coordinates.

        - To acquire data for autonomous navigation:
            - To memorize waypoints coordinates.
            - To calculate flight coordinates.
    - To acquire climb, descent and approach trajectory.
    - To acquire landing path.
    - To acquire different approach trajectory.
    - To acquire missed approach procedure.
    - To acquire holding procedure.
    - To acquire waypoints altitude.
    - To memorize flight plan.
- Data processing:
    - To calculate optimal trajectory for all flight segment (climb, cruise, descent).
    - To calculate trajectory in case of critical failure.
    - To calculate flight speed for each segment of mission profile.
    - To calculate heading, attitude, distance and time to reach each waypoints when they are VOR-DME station.
    - To calculate heading, attitude, distance and time to reach each waypoints when they are not VOR-DME station.
    - To determine lateral and vertical deviation of actual trajectory in comparison to the proper one for climb, cruise and descent segments.
    - To determine lateral and vertical deviation of actual trajectory in comparison to the proper one for approach and landing segments.
    - To provide surrounding terrain altitude.
- Data management:
    - To manage waypoints database.
    - To manage airports database.
    - To store and update navigation data.
    - To store and update weather data.
    - To verify acquired and calculated data accuracy.

- To display information:
    - To display flight route and waypoints.
    - To provide visual and acoustic warning in case of traffic collision.
    - To display heading.
    - To display true heading.
    - To display environment data.
    - To provide visual and acoustic warning in case of potential ground collision.
    - To display surrounding terrain altitude.
    - To display weather situation.
    - To display approach and landing correct trajectory.
    - To display trajectory in case of missed approach.

On the basis of the basic functions listed above, the functions/devices matrix can be created which, for sake of simplicity, illustrates only part of the complete functions/devices matrix related to the top level function "to allow navigation". It is worth remembering that, as in this case the Functional Analysis is applied at subsystem level, the basic components are the main subsystem equipments. The functions/devices matrix has thus been called functions/equipments matrix.

Functional Analysis at System Level: The Cubesat e-st@r

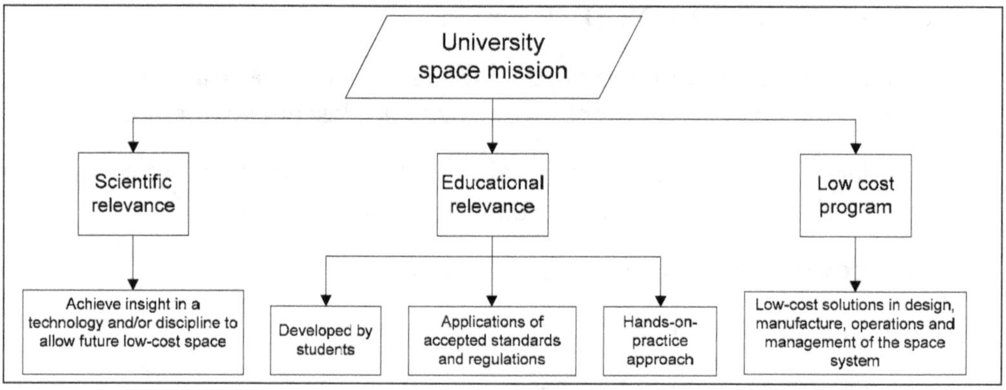

e-st@r program guidelines.

The e-st@r (Educational SaTellite @ politecnico di toRino) program is taken as case-study. The project is an educational initiative carried out by students and researchers of Politecnico di Torino within an ESA program aiming at the launch and orbit operations of nine cubesats, developed by as many European Universities, to promote space activities among young generations. E-st@r program guidelines are illustrated in figure.

The mission statement sounds as follows: "Educate aerospace-engineering students on complex systems design and management, team work, and standards implementation. Achieve insight in the development of enabling technologies for low-cost access to space".

The following assumptions can be derived from the mission statement:

- The program shall be carried out by students. They shall design, manufacture, verify and test, and operate a space system.
- The program shall be carried out in compliance with current regulations and applicable standards.
- The program shall have educational relevance, which means that students must learn by practice.
- An experiment shall be included in the space system. The experiment shall be simple and cheap, but at the same time it must permit to achieve insight in a discipline and technology to be used in the future to allow low-cost space mission.
- The program driver shall be the research for low-cost solutions in design, manufacture, operations and management of space systems.

Notwithstanding the necessity of keeping cost down and taking into account the educational spirit of the e-st@r program, which implies the will of enhancing the interests and competences of the students, e-st@r has also scientific objectives, which reflect real interests of the scientific and industrial communities. Taking into account all high-level requirements and constraints, as a result of a trade-off analysis it has been decided that the system would accomplish a mission aimed at testing an active Attitude Determination and Control System (ADCS).

The cubesat shall be piggybacked by the Vega LV during its Maiden Flight. Mission duration for this kind of project shall be in the range of 3-12 months. The cubesat shall be operated from ground in a simply and cheap way. High grade of operations autonomy is desirable. Students shall be designers, developers, manufacturers, operators and managers of the entire mission. The mission shall demonstrate some kind of non-space technologies and try to space-qualify them. The primary payload shall be a simple active ADCS. As secondary payload, the test of commercial items is considered. The mission data shall be available to the cubesat community and to radio-amateurs union. No commercial purposes shall be pursued.

Functional Analysis methodology has been used to derive the requirements for the system and to determine which subsystems are needed to carry out the mission. The second iteration of the Functional Analysis allows deriving next level requirements for equipments and components. A part of the complete functional tree for the e-st@r mission is shown in figure. The mission segments are identified by the first level functions (i.e. "To connect ground and space segments", "To do on ground operations", "To reach the orbit", "To do in orbit operations" and "To comply with space debris mitigation regulations") and they reflect the mission architecture's elements.

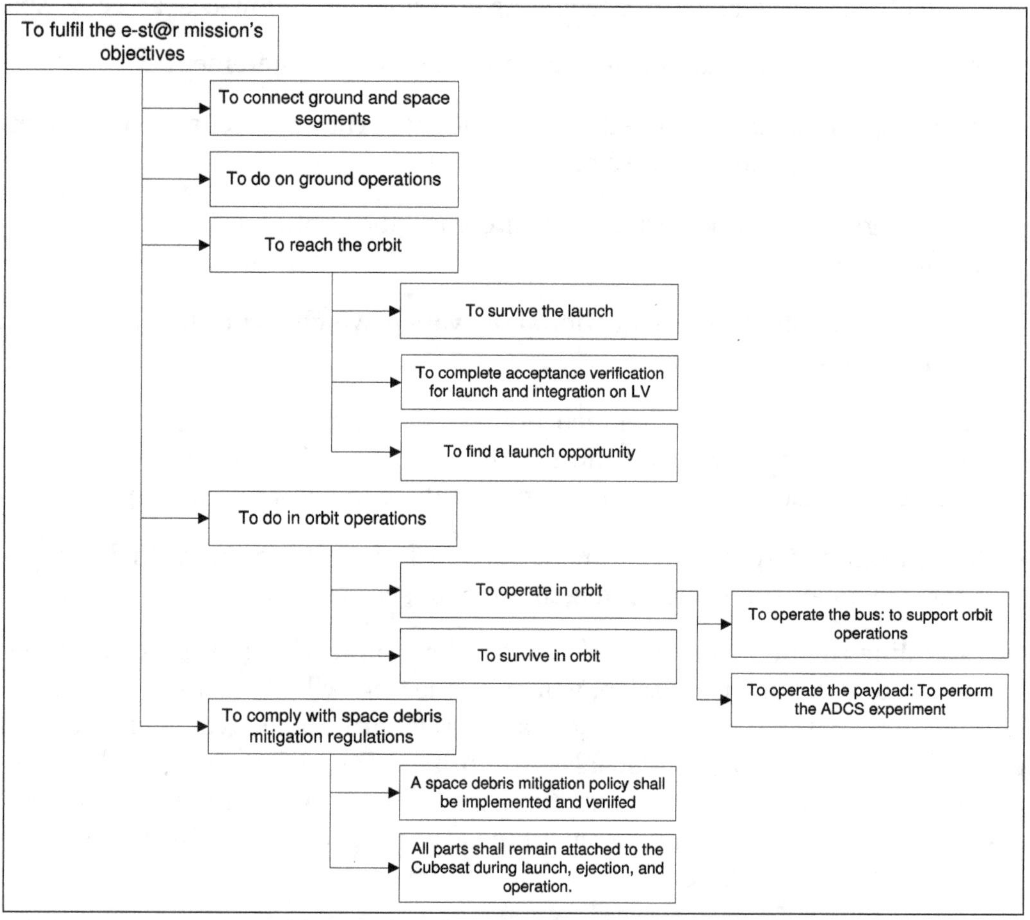

Part of the complete functional tree for e-st@r mission.

The elements of the e-st@r mission architecture are reported hereafter:

- Space segment: Cubesat = payload and bus.
- Ground segment: One main ground control station + one backup ground control station (mobile and transportable). Radio amateur network. Cubesat laboratory at Polito.
- Launch segment: Vega LV and CSG (French Guyana).
- Subject: Data measurement.
- Orbit: Direct injection into LEO (approx 300x1450km, 70°).
- Operations: Students at main and backup GCSs. Data processing at Cubesat lab for deeper investigation, or in case of emergency.
- Communications.

As an example, the product tree of two elements of the e-st@r mission architecture, the space (i.e. the cubesat, made up of payload and bus) and the ground segment, is shown in figure. It is worth noting that, while the space segment can be directly linked through a functions/devices matrix to the first level function "To do in orbit operations", the ground segment can be directly linked to the first level function "To do ground operation". Eventually figure illustrates the physical block diagram of the cubesat. The block diagram shows all subsystems (apart from structures) and their connections. The design and sizing of the subsystems in phase A have been carried out using common available methods.

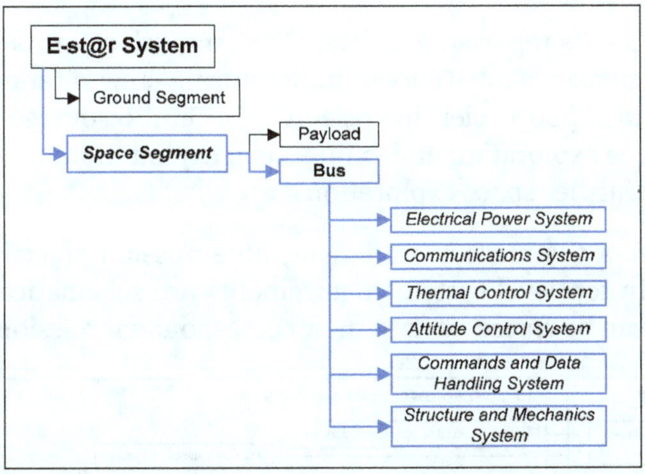

Product tree of the e-st@r system: The ground and the space segment.

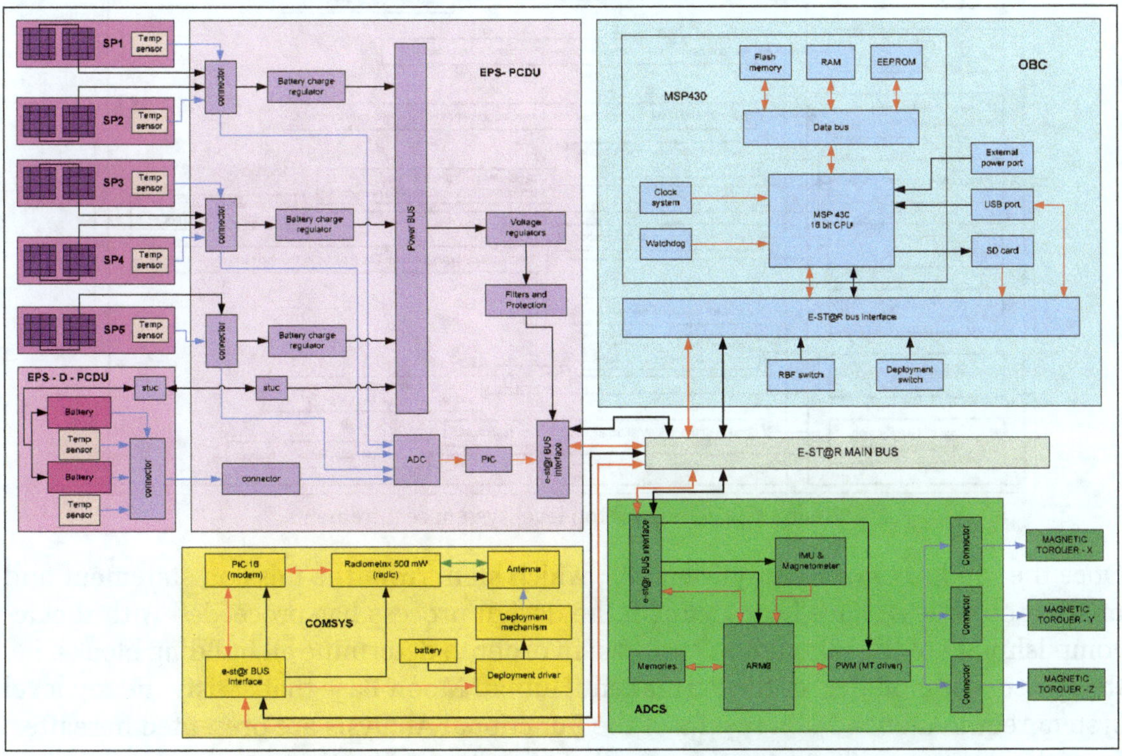

Physical block diagram of the e-st@r cubesat.

Functional Analysis at System of Systems Level: The Permanent Human Moon Base PHOEBE

The system of systems here considered is a permanent human Moon base. The Functional Analysis methodology has been applied, in order to accomplish the primary objectives, i.e. in order to develop the functional tree and the product tree of the Moon base.

The Moon base has been given the name PHOEBE, which stays for: Permanent Human moon Exploration Base.

The mission statement is reported hereafter: "To establish a permanent lunar base for a nominal crew of 18 astronauts (maximum 24 during crew rotation) with a turnover time of 6 months, to support scientific research, In-Situ Resources Utilization (ISRU) development, surface exploration and commercial exploitation; its evolution will provide an outpost for further space exploration".

After the definition of the mission statement, nine mission objectives have been determined. The main top level system requirements are schematically represented in figure, where they can be traced back to their correspondent mission objectives.

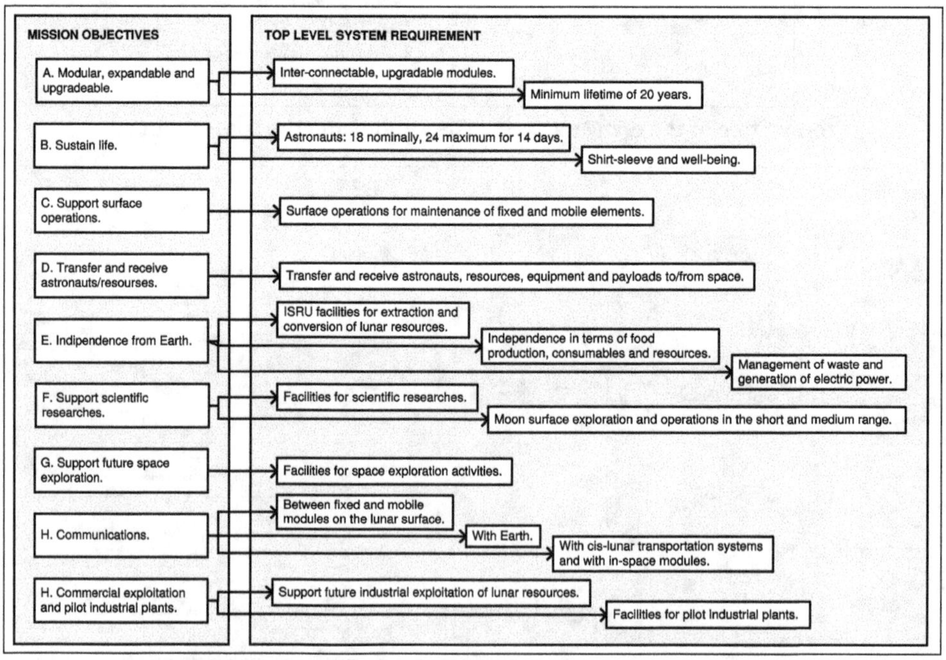

Mission objectives and top level system requirements.

Once the top level system requirements, which stem from the mission statement and mission objectives, have been defined, the design process has proceeded with the accomplishment of the Functional Analysis, in order to determine all building blocks, i.e. the systems or modules, of the Permanent Human Moon Base that satisfy the top level system requirements. Main results of the Functional Analysis are presented hereafter. In particular figure illustrates the so-called "first level" functional tree, where the top level function "To carry out a Permanent Human Moon Base" has been split into 10 first level functions. Each first level function has then been divided into lower level functions to identify the basic level functions, i.e. those functions that can immediately be connected to one building block of the Moon base.

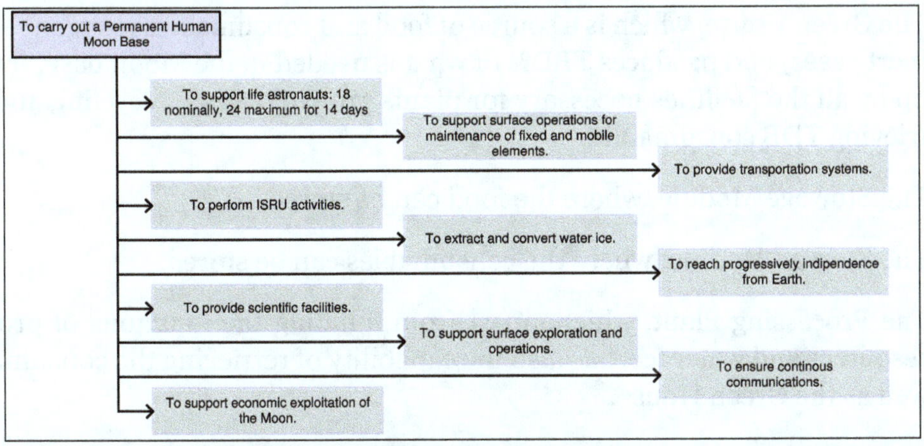

PHOEBE First level functional tree.

For sake of clarity, figure shows how the first level function "To reach progressively independence from Earth" has been decomposed into its basic functions:

- To provide plants growth facilities.
- To store the food produced.
- To extract resources from the waste.
- To retrieve TBD (To Be Defined) consumables.
- To store the retrieved consumable.
- To provide the electrical power.

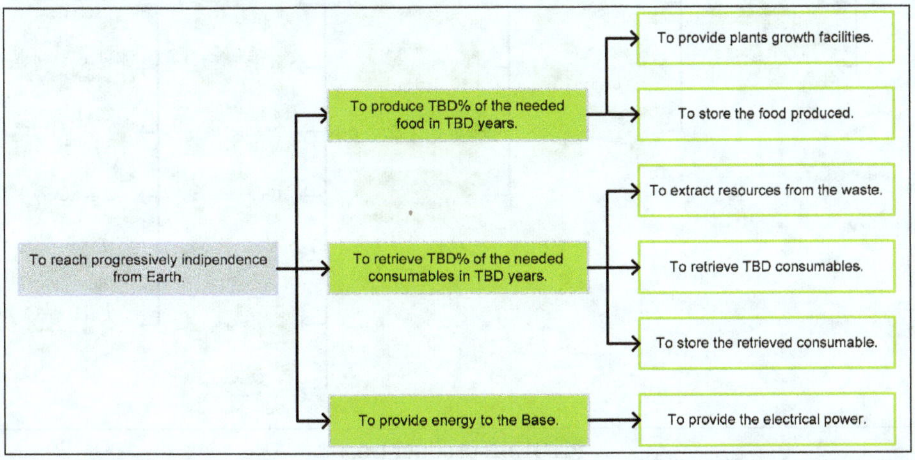

Functional tree of the first level function: "To reach progressively independence from Earth".

Once the basic functions have been identified, it is possible to choose the building blocks of the Moon base that will perform those functions. Considering for instance the basic functions presented in figure, the corresponding building blocks can be obtained through the functions/building blocks (or functions/devices) matrix.

- The Green House, which is a source of food and consumables (oxygen recycling, inert gases) and produces TBD% of what is needed in the Moon base. It is made up by all the facilities necessary for plants growing, waste recycling and for retrieving TDB consumables.
- The Stowage Module, where the food can be stored.
- The Storage Module, where the consumables can be stored.
- The Processing Plant, which, apart from fulfilling the functions of processing resources and water ice, has also the capability of retrieving the consumables as well as the Green House.
- The Material Science Laboratory, which, apart from fulfilling the functions of processing resources and water ice and providing materials experiments and test facilities for space exploration technology, has also the capability of retrieving the consumables as well as the Green House.
- The Power Plant, which provides the Moon Base with electric power.

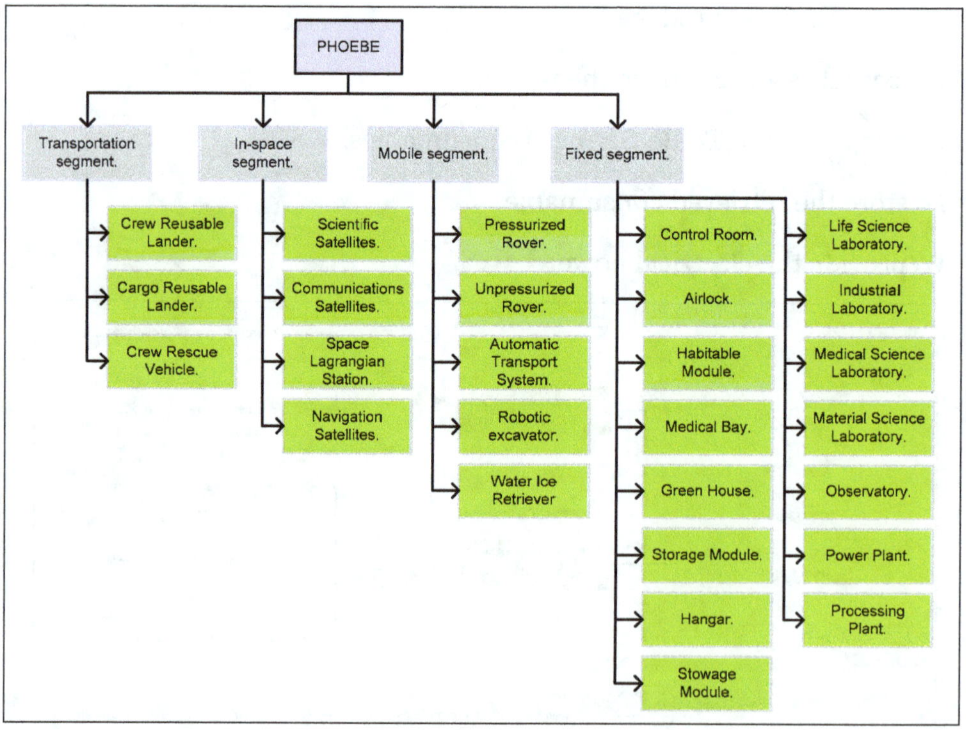

PHOEBE product tree.

Applying the same methodology to all the other first level functions listed in figure, the complete product tree of the Moon base, i.e. all PHOEBE systems, can be obtained, as figure illustrates, where the various building blocks have been grouped into four different categories or segment: transportation, in-space, mobile and fixed segment.

The identification of all systems of the Moon base and the understanding of their related functions are the final results of the presented Functional Analysis at system of systems level.

System Integration

System Integration is the process of integrating all the physical and virtual components of an organisation's system. The physical components consist of the various machine systems, computer hardware, inventory, etc. The virtual components consist of data stored in databases, software and applications. The process of integrating all these components, so that act likes a single system, is the main focus of system integration.

The main reason for organizations to use system integration is their need to improve productivity and quality of their operations. The goal is to get the organizations various IT systems to "talk to each other" through the integration, to speed up information flows and reduce operational costs for the organization. But system integration is not used only to connect an organization's internal systems, but also third parties that the organization operates with.

System Integration Methods

Typical System integration methods are divided into the following different categories:

Point-to-point Integration

One could argue that a point-to-point integration (or point-to-point connection) is not a system integration as such since there are only two system components involved. However, while it lacks the complexity of "true" system integration, it still connects a system to another system for them to function together. Typically, such point-to-point integration handles only one function and does not involve any complex business logic. Many cloud-based applications offer these types of point-to-point integrations as productized, "out of the box" integration modules for the most common IT systems.

Vertical Integration

In vertical integration method, the system components (sub-systems) are integrated by creating functional "silos", beginning with the basic bottom function upward. This is normally relatively simple and easy method that only involves a limited number of systems (more than two), but on the other hand, this integration method is quote rigid and more difficult to manage in the long term as any new functionally will require its own functional "silo". Still, this method can be used effectively to create simple integrations that only need to address a single function.

Star Integration

Star integration means that a system where each sub-system is connected with other sub-systems using point-to-point connections. This allows for more functionality, but as the number of integrated systems increases the number of integrations also increases significantly, and the management of the integrations becomes very demanding. As an example, to connect ten systems to each other using this method, will require 45 separate integrations, and every time there is a change in one system, nine connections may need to be re-done as well. Sometimes Star Integration is there for also referred to as "Spaghetti Integration" as an analogy to "Spaghetti code".

Horizontal Integration

In horizontal integration, a separate sub-system is used as a common interface layer between all sub-systems. Very often this layer is referred to as an Enterprise Service Bus (ESB). This method allows each sub-system to have just one single interface to communicate with all the other sub-systems connected to the common interface layer (i.e., with ten system, there are only ten connections). The benefit of this method is also that each sub-system can be changed or even replaced without having to re-do the interfaces of any other systems.

Common Data Format Integration

Integration different IT systems to each other usually require that the data coming out from one system needs to be transformed to a different data format used by the receiving system. As with the Star Integration, if each transformation needs to be done system-by-system basis, the number of data transformations increases significantly and becomes a high maintenance task. To overcome this problem, common data format approach, allows each system to do only one data conversion from its native format to the common (and vice versa). This way the number of required data transformations is just a high as the number of the sub-system.

What is the System Integrator's Role?

On broad terms in the IT world, a Systems Integrator ("SI") is regarded as a company that specializes in implementing, planning, coordinating, scheduling, testing, improving and sometimes maintaining IT systems. Good examples of SIs are, e.g., Deloitte, IBM, Accenture, TCS, etc. They deliver large IT projects (e.g., ERP projects) trying to manage such projects and the numerous suppliers involved. However, in terms of system integration, the role of the systems integrator is narrowed down to enabling the data integrations between the different existing systems of the end customer defined in the project scope. This may mean anything starting from simple internal point-to-point connections to very complex many-to-many integrations both internally and with third parties.

The systems integrators role in this equation is normally to design, implement and test the integration solution, but the role of the systems integrator may also include continuous management of the solutions as well as contacting third parties to enable connections with them. Most importantly, however, the systems integrator brings to the table integration expertise that the customer is otherwise lacking internally (or has a shortage of available internal resources at hand).

Why is B2B Integration more Relevant than Ever?

Business-to-Business integration is by no means a new concept. Some IT-driven companies started implementing B2B Integration projects almost 50 years ago (and hopefully by now some of them have even completed these projects). B2B Integration basically means the integration, automation, and optimization of business processes that extend outside the company firewall. While these processes may vary significantly between them, the one common thing that they share is that the integration of such external business processes delivers sustainable competitive advantage to the organization. Such advantages may include, e.g., real-time visibility, improved automation, inventory optimization, and improved customer satisfaction.

Companies have recognized that having good software solutions is simply not enough. They may be using the most functionally-rich software applications within their own firewall (or in the cloud), but without proper B2B connectivity and associated capabilities, they cannot effectively manage, e.g., their end-to-end supply chain process.

While B2B Integration initially began with large enterprises mandating methods of receiving business information, it quite quickly evolved into the adoption of Electronic Data Interchange (EDI) standards and later on to other newer technologies such as XML, JSON, etc. Nowadays it seems that every new application has some type of API that allows integration with such application. However, this leaves the task of actually integrating such API with other systems, and more often than enough most companies simply do not know how to make this happen.

System Integration Challenges

System integration is not rocket science by any means, but there are a number of issues that make system integration challenging for all organizations. According to certain studies, up to 70% of all integration projects fail in some aspect. This is probably more common for complex projects, but sometimes even relatively simple system integration projects can run into trouble. Most of the failures are not due to the chosen integration technology itself or technical difficulties with the systems in the scope, but due to project and change management issues.

Typical reasons for system integration project failure include for example:

Constant Changes of the Integration Landscape

The longer the project takes, the more significant this issue becomes. To manage this risk, time is of the essence, keeping the integration projects short improves the success rate of the project. Further, an agile working methodology that can cater to changing requirements along the way and also after the project is essential for the success of the systems integration.

Lack of Skilled Resources

System integration requires expertise that is not easy to come by. Having excellent integration technology is not enough if the required expertise is not there. Most companies struggle to find and retain employees with the required skills set for system integration. The best way to tackle this issue is to use an external third-party provider that can bring the needed integration expertise into the table as required, in addition to providing the integration technology.

Lack of Accountability

When you're integrating many different sub-systems, the accountability for the success of the integration becomes blurred very easily. You may have multiple stakeholders (e.g., vendors, system owners, etc.) in the equation, none of whom is responsible for the entire system integration. They only care and handle their side of integration at best, but they will not venture outside their own territory. But integration always has more than one party. So, when something goes wrong, the situation turns very easily to finger pointing and blaming the other parties, instead of someone "owning" the integration. If a single party handles the system integration project, that party is also (often contractually) responsible for the success of such a system integration project, and there is no ambiguity over accountability.

Legacy System Integration

Most companies that have done business for decades have some old, legacy IT systems in use and running on their own on-premise servers. These systems may be essential to the core business of the organization and cannot be replaced easily with a more modern IT system. Integration with such legacy systems may be tricky as they may completely lack any ready-made interfacing capability. Most systems do, however, have the capability to read or write information into a file folder that can the accessed, e.g., through an FTP connection by another system, but sometimes the only way to integrate such sub-systems with other sub-systems is to read and/or write data directly in their database.

A modern integration solution must be able to handle also these type of integration scenarios. Cloud-based iPaaS solutions typically use local, on-premise adapters that provide the needed functionality for these integrations. Such adapters act as an active local "interface" between the passive legacy system (or its database) and the cloud-based iPaaS solution. If needed, additional business rules and other functionality regarding the legacy system integration will be handled in the iPaaS service, keeping the maintenance of such business logic centralized and easy. This way the customer does not need to make any costly changes in its legacy IT systems, but the system integrator can provide the integration logic outside the company firewall.

Cognitive Systems Engineering

Cognitive systems engineering is a specialty discipline of systems development that addresses the design of socio-technical systems. A sociotechnical system is one in which humans provide essential functionality related to deciding, planning, collaborating and managing. Drawing on contemporary insights from cognitive, social and organizational psychology, cognitive systems engineers seeks to design systems that are effective and robust. The focus is on amplifying the human capability to perform cognitive work by integrating technical functions with the human cognitive processes they need to support and on making that cognitive work more reliable.

Cognitive systems engineers assist with the design of human interfaces, communication systems, training systems, teams, and management systems. They employ principles and methods that bear on the design of procedures, processes, training and technology. Examples of systems that can benefit are military command and control, civil air traffic control, transportation, communication, process control, power generation, power distribution, health-care, and large-scale project infrastructure.

Role for Cognitive Systems Engineering

The need for a systematic and comprehensive approach to cognitive issues in the design of sociotechnical systems has emerged over the past 20 years as computer-based technologies have pushed the nature of operational work in a direction in which cognitive challenges dominate. Issues such as decision-making in complex and dynamic information environments, distributed collaboration, and management of extensively networked systems have, in many cases, transformed the nature of work. Cognitive systems engineers identify the cognitive states, the cognitive processes, and the cognitive strategies used by skilled practitioners to perform this work and subsequently develop design solutions for such things as decision and planning tools that support expert human cognition.

At an abstract level of description, the approach taken is straightforward and not all that different to the analysis and design strategies used in other areas of engineering.

Cognitive systems engineers observe the field practice and represent the knowledge thus acquired in some form that facilitates the design of appropriate cognitive support systems. Those design solutions are then evaluated via computer modelling or human-in-the-loop simulation. While this approach will sound familiar to many engineers, the tools used for knowledge acquisition, knowledge representation and cognitive modelling will not. These have been developed specifically to deal with the complex and nonlinear nature of human cognition; its hidden interdependencies and those of its processes that are beyond the conscious awareness of the operational expert.

Within the development of any large-scale, sociotechnical system, cognitive systems engineering has important roles to play throughout the acquisition life cycle. It has tools and strategies that can be used to identify human-relevant operational requirements and operational demands, to generate human-compatible solution descriptions, and to design cognitive support systems. Finally, it has tools for verifying and validating cognitive performance and for monitoring and enhancing system safety performance.

References

- Classification-of-systems, signals-systems: ecetutorials.com, Retrieved 14 May, 2020
- Properties-of-a-system: myclassbook.org, Retrieved 27 July, 2020
- Systems-engineering, topic: britannica.com, Retrieved 18 January, 2020
- Applications-of-systems-engineering, systems-engineering-topic: britannica.com, Retrieved 02 March, 2020
- Modern system analysis and design, fourth edition, Pearson Prentice Hall, New Jersey, 2002
- Models of expertise as blueprints cognitive engineering: Applications to landmine detection. Proceedings of the 48th Annual Meeting of the Human Factors and Ergonomics Society, 48, 458–462

2
Systems Architecture

A type of conceptual model that deals with the organization and management of structure and behavior of a system is termed as systems architecture. Architecture description language, lifecycle modeling language, etc. are a few concepts that fall under its domain. This is an introductory chapter which will briefly introduce about systems architecture.

The purpose of system architecture activities is to define a comprehensive solution based on principles, concepts, and properties logically related and consistent with each other. The solution architecture has features, properties, and characteristics satisfying, as far as possible, the problem or opportunity expressed by a set of system requirements (traceable to mission/business and stakeholder requirements) and life cycle concepts (e.g., operational, support) and are implementable through technologies (e.g., mechanics, electronics, hydraulics, software, services, procedures, human activity).

System Architecture is abstract, conceptualization-oriented, global, and focused to achieve the mission and life cycle concepts of the system. It also focuses on high-level structure in systems and system elements. It addresses the architectural principles, concepts, properties, and characteristics of the system-of-interest. It may also be applied to more than one system, in some cases forming the common structure, pattern, and set of requirements for classes or families of similar or related systems.

General Concepts and Principles

Notion of Structure

The SEBoK considers systems engineering to cover all aspects of the creation of a system, including system architecture.

The majority of interpretations of system architecture are based on the fairly intangible notion of structure (i.e. relationships between elements). Some authors limit the types of structure considered to be architectural; for example, restricting themselves to functional and physical structure. Recent practice has extended consideration to include behavioral, temporal and other dimensions of structure.

Architecture description provides a useful description of the architecture considering the stakeholder concerns, architecture viewpoints, architecture views, architecture models, architecture descriptions, and architecting throughout the life cycle. An attempt to develop and apply a systematic approach to characterizing architecture belief systems in systems engineering has been described by the INCOSE UK Architecture Working Group.

Architecture Description of the System

An architecture framework contains standardized viewpoints, view templates, meta-models, model templates, etc. that facilitate the development of the views of a system architecture. ISO/IEC/IEEE 42010 specifies the normative features of architecture frameworks, viewpoints, and views as they pertain to architecture description. A viewpoint addresses a particular stakeholder concern (or set of closely related concerns). The viewpoint specifies the kinds of model to be used in developing the system architecture to address that concern (or set of concerns), the ways in which the models should be generated, and how the models are related and used to compose a view.

Logical and physical models (or views) are often used for representing fundamental aspects of the system architecture. Other complementary viewpoints and views are necessarily used to represent how the system architecture addresses stakeholder concerns, for example, cost models, process models, rule models, ontological models, belief models, project models, capability models, data models, etc.

Classification of Principles and Heuristics

Engineers and architects use a mixture of mathematical principles and heuristics (heuristics are lessons learned through experience, but not mathematically proven). When an issue is identified and defined through system requirements, principles and heuristics may or may not be able to address it. Principles and heuristics that are used in system views/models can be classified according to the domains in which those system views/models are used, as follows:

- Static domain relates to physical structure or organization of the SoI broken down into systems and system elements. It deals with partitioning systems, system elements, and physical interfaces.
- Dynamic domain relates to logical architecture models; in particular, to the representation of the behavior of the system. It includes a description of functions (i.e. transformations of input flows into output flows) and interactions between functions of the system and between those of the external objects or systems. It takes into account reactions to events that launch or stop the execution of functions of the system. It also deals with the effectiveness (i.e. performances, operational conditions) of the system.

- Temporal domain relates to temporal invariance levels of the execution of functions of the system. This means that every function is executed according to cyclic or synchronous characteristics. It includes decisional levels that are asynchronous characteristics of the behavior of some functions.
- Environmental domain relates to enablers (production, logistics support, etc.), but also to the survivability of the system in reaction to natural hazards or threats and to the integrity of the system in reaction to internal potential hazards. This includes, for example, climatic, mechanical, electromagnetic, and biological aspects.

Transition from System Requirements to Logical and Physical Architecture Models

The aim of the approach is to progress from system requirements (representing the problem from a supplier/designer point of view, as independent of technology as possible) through an intermediate model of logical architecture, to allocate the elements of the logical architecture model to system elements of candidate physical architecture models. (System requirements and logical architecture models share many characteristics, as they are both organized on functional lines, independently of the implementation. Some authors go so far as to conflate the two, which simplifies the handling of multiple simultaneous views. Whether this approach is adopted depends on the specific practices of the development organization and where contractual boundaries are drawn).

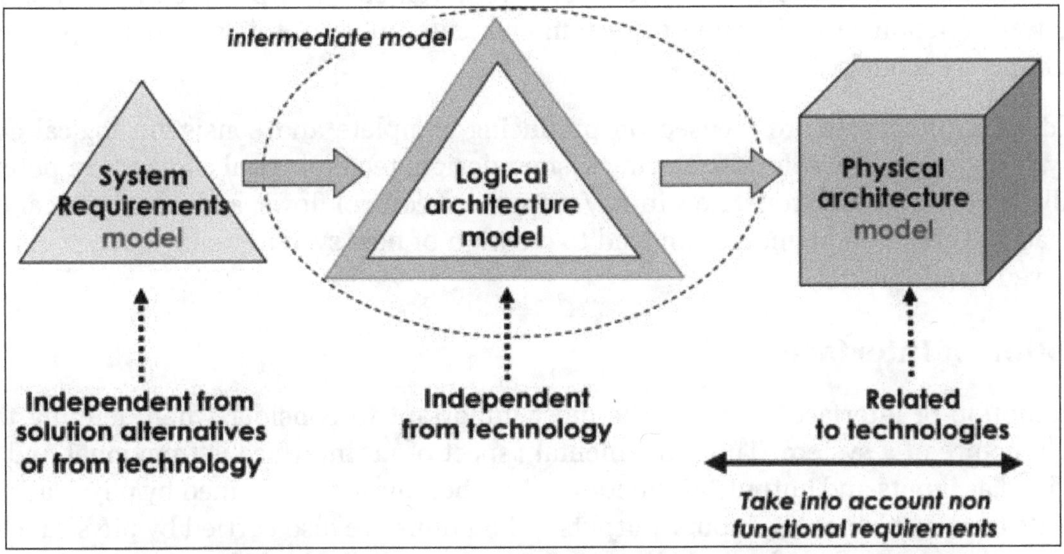

Usage of Intermediate Logical Architecture Models during Architecture and Design.

Design decisions and technological solutions are selected according to performance criteria and non-functional requirements, such as operational conditions and life cycle constraints (e.g., environmental conditions, maintenance constraints, realization constraints, etc.), as illustrated in figure. Creating intermediate models, such as logical architecture models, facilitates the validation of functional, behavioral, and temporal properties of the system against the system requirements that have no major technological influence impacts during the life of the system, the physical interfaces, or the technological layer without completely questioning the logical functioning of the system.

Iterations between Logical and Physical Architecture Model Development

In system requirements, the exact approach taken in the synthesis of solutions will often depend on whether the system is an evolution of an already understood product or service, or a new and unprecedented solution.

Whatever the approach, architecture activities require spending several iterations between logical architecture models development and physical architecture models development, until both logical and physical architecture models are consistent and provide the necessary level of detail. One of the first architecture activities is the creation of a logical architecture model based on nominal scenarios (of functions). The physical architecture model is used to determine main system elements that could perform system functions and to organize them.

Subsequent logical architecture model iterations can take into account allocations of functions to system elements and derived functions coming from physical solution choices. It also supplements the initial logical architecture model by introducing other scenarios, failure analyses, and operational requirements not previously considered. Derived functions are allocated to system elements; in turn, this affects the physical architecture models.

Additional iterations are focused on producing complete and consistent logical and physical views of the solution. During system design, technological choices can potentially lead to new functions, new input/output and control flows, and new physical interfaces. These new elements can lead to creation of new system requirements, called derived requirements.

Notion of Interface

The notion of interface is one of the most important to consider when defining the architecture of a system. The fundamental aspect of an interface is functional and is defined as inputs and outputs of functions. As functions are performed by physical elements (system elements), inputs/outputs of functions are also carried by physical elements; these are called physical interfaces. Consequentially, both functional and physical aspects are considered in the notion of interface. A detailed analysis of an interface shows the function "send" located in one system element, the function "receive" located in the other one, and the function "carry" as being performed by the physical interface that supports the input/output flow.

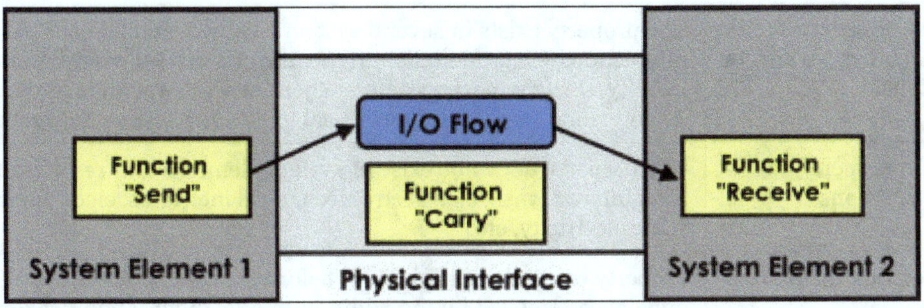

Complete Interface Representation.

In the context of complex exchanges between system elements, particularly in software-intensive systems, a protocol is seen as a physical interface that carries exchanges of data. However, the input/output flows can include many other exchanges than data, such as energy.

Emergent Properties

The overarching architecture of a system may have design properties or operational effects that emerge from the arrangement and interaction between system elements, but which may not be properties of any individual element or intended for the system as a whole.

The elements of an engineered system interact among themselves and can create desirable or undesirable phenomena, such as inhibition, interference, resonance, or the reinforcement of any property. The definition of the system includes an analysis of interactions between system elements in order to prevent undesirable properties and reinforce desirable ones.

A property which emerges from a system can have various origins, from a single system element to the interactions among several elements. The term emergent properties is used by some authors to identify any property which emerges from a system, while other may refer to this as synergy and reserve emergent property for explaining unexpected properties or properties not considered fully during system development, but have emerged during operation.

Table: Properties and Emergent Properties.

Broad Categories of Properties	Description and Examples
Local Property	The property is located in a single system element – e.g. the capacity of a container is the capacity of the system.
Accumulative System Property	The property is located in several system elements and is obtained through the simple summation of elemental properties – e.g. the weight of the system results from the sum of the weights of its system elements.

Emergent Property Modified by Architecture and/or Interactions.	The property exists in several system elements and is modified by their interactions – e.g. the reliability/safety of a system results from the reliability/safety of each system element and the way they are organized. Architectural steps are often critical to meeting system requirements.
Emergent Property Created by Interactions	The property does not exist in system elements and results only from their interactions – e.g. electromechanical interfaces, electromagnetism, static electricity, etc.
Controlled Emergent Property	Property controlled or inhibited before going outside the system – e.g.: unbalance removed by the addition of a load; vibration deadened by a damper.

Physical architecture design will include the identification of likely synergies and emergent properties and the inclusion of derived functions, components, arrangements, and/or environmental constraints in the logical or physical architectures models to avoid, mitigate or restrain them within acceptable limits. Corresponding derived requirements should be added to the system requirements baseline when they impact the system-of-interest(SoI). This may be achieved through the knowledge and experience of the systems engineer or through the application of system patterns. However, it is generally not possible to predict, avoid, or control all emergent properties during the architecture development. Fully dealing with the consequences of emergence can only be done via iteration between system definition, system realization and system deployment and use.

The notion of emergence is applied during architecture and design to highlight necessary derived functions; additionally, internal emergence is often linked to the notion of complexity. This is the case with complex adaptive systems (CAS), in which the individual elements act independently, but behave jointly according to common constraints and goals. Examples of CAS include: the global macroeconomic network within a country or group of countries, stock market, complex web of cross border holding companies, manufacturing businesses, geopolitical organizations, etc.

Reuse of System Elements

Systems engineers frequently utilize existing system elements. This reuse constraint has to be identified as a system requirement and carefully taken into account during architecture and design. One can distinguish three general cases involving system element reuse.

Table: System Element Re-use Cases.

Re-use Case	Actions and Comments
Case 1: The requirements of the system element are up-to-date and it will be re-used with no modification required.	• The system architecture to be defined will have to adapt to the boundaries, interfaces, functions, effectiveness, and behavior of the re-used system element. • If the system element is not adapted, it is probable that costs, complexity, and risks will increase.

Case 2: The requirements of the system element are up-to-date and it will be re-used with possible modifications.	• The system architecture to be defined is flexible enough to accommodate the boundaries, interfaces, functions, effectiveness, and behavior of the re-used system element. • The design of the reused system element, including its test reports and other documentation, will be evaluated and potentially redesigned.
Case 3: The requirements are not up-to-date or do not exist.	• It is necessary to reverse engineer the system element to identify its boundaries, interfaces, functions, performances, and behavior. • This is a difficult activity, since the extant documentation for the re-used system element is likely unavailable or insufficient. • Reverse engineering is expensive in terms of both time and money, and brings with it increased risk.

There is a common idea that reuse is free; however, if not approached correctly, reuse may introduce risks that can be significant for the project (costs, deadlines, complexity).

Process Approach

Purpose

The purpose of the System Architecture process is to generate system architecture alternatives, to select one or more alternative(s) that frame stakeholder concerns and meet system requirements, and to express this in a set of consistent views. It should be noted that the architecture activities below overlap with both system definition and concept definition activities. In particular that key aspects of the operational and business context, and hence certain stakeholder needs, strongly influence the approach taken to architecture development and description. Also, the architecture activities will drive the selection of, and fit within, whatever approach to solution synthesis has been selected.

Activities of the Process

Major activities and tasks performed during this process include the following:

Initialize the Definition of the System Architecture

- Build an understanding of the environment/context of use for which a system is needed in order to establish insight into the stakeholder concerns. To do this, analyse relevant market, industry, stakeholder, enterprise, business, operations, mission, legal and other information that help to understand the perspectives that could guide the definition of the system architecture views and models.

- Capture stakeholder concerns (i.e., expectations or constraints) that span system life cycle stages. The concerns are often related to critical characteristics to the system that relate to the stages; they should be translated into or incorporated to system requirements.
- Tag system requirements that deal with operational conditions (e.g., safety, security, dependability, human factors, interfaces, environmental conditions) and life cycle constraints (e.g., maintenance, disposal, and deployment) that would influence the definition of the architecture elements.
- Establish an architecture roadmap and strategy that should include methods, modeling techniques, tools, need for any enabling systems, products, or services, process requirements (e.g., measurement approach and methods), evaluation process (e.g., reviews and criteria).
- Plan enabling products or services acquisition (need, requirements, procurement).

Necessary Architecture Viewpoints

Based on the identified stakeholder concerns, identify relevant architecture viewpoints and architecture frameworks that may support the development of models and views.

Develop Candidate Architectures Models and Views

- Using relevant modeling techniques and tools, and in conjunction with the Stakeholder Needs and Requirements process and the System Requirements process, determine the system-of-interest context including boundary with elements of the external environment. This task includes the identification of relationships, interfaces or connections, exchanges and interactions of the system-of-interest with external elements. This task enables to define or understand the expected operational scenarios and/or system behaviors within its context of use.
- Define architectural entities (e.g., functions, input/output flows, system elements, physical interfaces, architectural characteristics, information/data elements, containers, nodes, links, communication resources, etc.), which address the different types of system requirements (e.g., functional requirements, interface requirements, environmental requirements, operational conditions – dependability, human factors, etc., constraints – physical dimensions, production, maintenance, disposal).
- Relate architectural entities to concepts, properties, characteristics, behaviors, functions, and/or constraints that are relevant to decisions of the system-of-interest architecture. This gives rise to architectural characteristics (e.g., generality, modularity, operability, efficiency, simplicity).

- Select, adapt, or develop models of the candidate architectures of the system, such as logical and physical models. It is sometimes neither necessary nor sufficient to use logical and physical models. The models to be used are those that best address key stakeholder concerns.
- From the models of the candidate architectures, compose views that are relevant to the stakeholder concerns and critical or important requirements.
- Define derived system requirements induced by necessary instances of architectural entities (e.g., functions, interfaces) and by structural dispositions (e.g., constraints, operational conditions). Use the system requirements definition process to define and formalize them.
- Check models and views consistency and resolve any identified issues. ISO/IEC/IEEE 42010, 2011 may be used for this.
- Verify and validate the models by execution or simulation, if modeling techniques and tools permit. Where possible, use design tools to check feasibility and validity; and/or implement partial mockups, or use executable architecture prototypes or simulators.

Relate System Architecture to System Design

- Define the system elements that reflect the architectural characteristics (when the architecture is intended to be design-agnostic, these system elements may be notional until the design evolves). To do this, partition, align, and allocate architectural characteristics and system requirements to system elements. Establish guiding principles for the system design and evolution. Sometimes, a "reference architecture" is created using these notional system elements as a means to convey architectural intent and to check for design feasibility.
- Define interfaces for those that are necessary for the level of detail and understanding of the architecture. This includes the internal interfaces between the system elements and the external interfaces with other systems.
- Determine the design properties applicable to system elements in order to satisfy the architectural characteristics.
- For each system element that composes the system, develop requirements corresponding to allocation, alignment, and partitioning of design properties and system requirements to system elements. To do this, use the stakeholder needs and requirements definition process and the system requirements definition process.

Assess Architecture Candidates and Select One

- Assess the candidate architectures using the architecture evaluation criteria.

This is done through application of the System Analysis, Measurement, and Risk Management processes.

- Select the preferred architecture(s). This is done through application of the Decision Management process.

Manage the Selected Architecture

- Establish and maintain the rationale for all selections among alternatives and decision for the architecture, architecture framework(s), viewpoints, kinds of models, and models of the architecture.
- Manage the maintenance and evolution of the architecture description, including the models, and views. This includes concordance, completeness, and changes due to environment or context changes, technological, implementation, and operational experiences. Allocation and traceability matrices are used to analyze impacts onto the architecture. The present process is performed at any time evolutions of the system occur.
- Establish a means for the governance of the architecture. Governance includes the roles, responsibilities, authorities, and other control functions.
- Coordinate reviews of the architecture to achieve stakeholder agreement. The stakeholder requirements and system requirements can serve as references.

Artifacts, Methods and Modeling Techniques

This process may create several artifacts, such as system architecture description documents and system justification documents (traceability matrices and architectural choices). The content, format, layout, and ownership of these artifacts may vary depending on the person creating them and the domains in which they are being used.

Practical Considerations

Pitfalls

Some of the key pitfalls encountered in planning and performing system architecture are provided in table.

Table: Pitfalls with System Architecture Definition.

Pitfall	Description
Problem Relevance	If the architecture is developed without input from the stakeholders' concerns, or cannot be understood and related back to their issues it might lose the investments of the stakeholder community.

Reuse of System Elements	In some projects, for industrial purposes, existing products or services are imposed very early as architecture/design constraints in the stakeholder requirements or in the system requirements, without paying sufficient attention to the new context of use of the system in which they are also included. It is better to work in the right direction from the beginning. Define the system first, taking note of other requirements, and then see if any suitable non-developmental items (NDI) are available. Do not impose a system element from the beginning, which would reduce the trade-space. The right reuse process consists of defining reusable system elements in every context of use.

Proven Practices

Some proven practices gathered from the references are provided in table.

Table: Proven practices with System Architecture Definition.

Practice	Description
Emerging properties	Control the emergent properties of the interactions between the systems or the system elements; obtain the required synergistic properties and control or avoid the undesirable behaviors (vibration, noise, instability, resonance, etc.).

Architecture Analysis and Design Language

The Architecture Analysis & Design Language (AADL) is an architecture description language standardized by SAE. AADL was first developed in the field of avionics, and was known formerly as the Avionics Architecture Description Language.

The Architecture Analysis & Design Language is derived from MetaH, an architecture description language made by the Advanced Technology Center of Honeywell. AADL is used to model the software and hardware architecture of an embedded, real-time system. Due to its emphasis on the embedded domain, AADL contains constructs for modeling both software and hardware components (with the hardware components named "execution platform" components within the standard). This architecture model can then be used either as a design documentation, for analyses (such as schedulability and flow control) or for code generation (of the software portion), like UML.

AADL Eco-system

AADL is defined by a core language that defines a single notation for both system and software aspects. Having a single model ease the analysis tools by having only one single representation of the system. The language specifies system-specific characteristics using properties.

The language can be extended with the following methods:

- User-defined properties: User can extend the set of applicable properties and add their own to specify their own requirements.
- Language annexes: The core language is enhanced by annex languages that enrich the architecture description. For now, the following annexes have been defined:
    - Behavior annex: Add components behavior with state machines.
    - Error-model annex: Specifies fault and propagation concerns.
    - ARINC653 annex: Defines modelling patterns for modelling avionics system.
    - Data-Model annex: Describes the modelling of specific data constraint with AADL.

AADL Tools

AADL is supported by a wide range of tools:

- OSATE that includes a modeling platform, a graphical viewer and a constraint query languages.
- Ocarina, an AADL toolchain for generating code from models.
- TASTE toolchain supported by the European Space Agency.

Related Projects

AADL has been used for the following research projects:

- AVSI/SAVI: An initiative that leverages AADL (among other languages) to perform virtual integration of aerospace and defense systems.
- META: A DARPA project for improving software engineering methods.
- PARSEC: A French initiative to validate and implement avionics systems from architecture models.
- TASTE: A platform for designing safety-critical systems from models.

Architecture Description Language

Architecture description languages (ADLs) are used in several disciplines: system engineering, software engineering, and enterprise modelling and engineering.

The system engineering community uses an architecture description language as a language and/or a conceptual model to describe and represent system architectures. The software engineering community uses an architecture description language as a computer language to create a description of a software architecture. In the case of a so-called technical architecture, the architecture must be communicated to software developers; a functional architecture is communicated to various stakeholders and users. Some ADLs that have been developed are: Acme (developed by CMU), AADL (standardized by the SAE), C2 (developed by UCI), SBC-ADL (developed by National Sun Yat-Sen University), Darwin (developed by Imperial College London), and Wright (developed by CMU).

The ISO/IEC/IEEE 42010 document, Systems and software engineering—Architecture description, defines an architecture description language as "any form of expression for use in architecture descriptions" and specifies minimum requirements on ADLs.

The enterprise modelling and engineering community have also developed architecture description languages catered for at the enterprise level. Examples include ArchiMate (now a standard of The Open Group), DEMO, ABACUS (developed by the University of Technology, Sydney). These languages do not necessarily refer to software components, etc. Most of them, however, refer to an application architecture as the architecture that is communicated to the software engineers.

Most of the writing below refers primarily to the perspective from the software engineering community.

A standard notation (ADL) for representing architectures helps promote mutual communication, the embodiment of early design decisions, and the creation of a transferable abstraction of a system. Architectures in the past were largely represented by box-and-line drawing annotated with such things as the nature of the component, properties, semantics of connections, and overall system behavior. ADLs result from a linguistic approach to the formal representation of architectures, and as such they address its shortcomings. Also important, sophisticated ADLs allow for early analysis and feasibility testing of architectural design decisions.

Characteristics

There is a large variety in ADLs developed by either academic or industrial groups. Many languages were not intended to be an ADL, but they turn out to be suitable for representing and analyzing an architecture. In principle ADLs differ from requirements languages, because ADLs are rooted in the solution space, whereas requirements describe problem spaces. They differ from programming languages, because ADLs do not bind architectural abstractions to specific point solutions. Modeling languages represent behaviors, where ADLs focus on representation of components. However, there are domain specific modeling languages (DSMLs) that focus on representation of components.

Minimal Requirements

The language must:

- Be suitable for communicating an architecture to all interested parties.
- Support the tasks of architecture creation, refinement and validation.
- Provide a basis for further implementation, so it must be able to add information to the ADL specification to enable the final system specification to be derived from the ADL.
- Provide the ability to represent most of the common architectural styles.
- Support analytical capabilities or provide quick generating prototype implementations.

ADLs have in common:

- Graphical syntax with often a textual form and a formally defined syntax and semantics.
- Features for modeling distributed systems.
- Little support for capturing design information, except through general purpose annotation mechanisms.
- Ability to represent hierarchical levels of detail including the creation of substructures by instantiating templates.

ADLs differ in their ability to:

- Handle real-time constructs, such as deadlines and task priorities, at the architectural level.
- Support the specification of different architectural styles. Few handle object oriented class inheritance or dynamic architectures.
- Support the analysis of the architecture.
- Handle different instantiations of the same architecture, in relation to product line architectures.

Positive Elements of ADL

- ADLs are a formal way of representing architecture.
- ADLs are intended to be both human and machine readable.
- ADLs support describing a system at a higher level than previously possible.

- ADLs permit analysis and assessment of architectures, for completeness, consistency, ambiguity, and performance.
- ADLs can support automatic generation of software systems.

Negative Elements of ADL

- There is no universal agreement on what ADLs should represent, particularly as regards the behavior of the architecture.
- Representations currently in use are relatively difficult to parse and are not supported by commercial tools.
- Most ADLs tend to be very vertically optimized toward a particular kind of analysis.

Common Concepts of Architecture

The ADL community generally agrees that Software Architecture is a set of components and the connections among them. But there are different kind of architectures like:

Object Connection Architecture

- Configuration consists of the interfaces and connections of an object-oriented system.
- Interfaces specify the features that must be provided by modules conforming to an interface.
- Connections represented by interfaces together with call graph.
- Conformance usually enforced by the programming language:
  - Decomposition: Associating interfaces with unique modules.
  - Interface conformance: Static checking of syntactic rules.
  - Communication integrity: Visibility between modules.

Interface Connection Architecture

- Expands the role of interfaces and connections:
  - Interfaces specify both "required" and "provided" features.
  - Connections are defined between "required" features and "provided" features.

- Consists of interfaces, connections and constraints.
  - Constraints restrict behavior of interfaces and connections in an architecture.
  - Constraints in an architecture map to requirements for a system.

Most ADLs implement an interface connection architecture.

Architecture vs. Design

Architecture, in the context of software systems, is roughly divided into categories, primarily software architecture, network architecture, and systems architecture. Within each of these categories, there is a tangible but fuzzy distinction between architecture and design. To draw this distinction as universally and clearly as possible, it is best to consider design as a noun rather than as a verb, so that the comparison is between two nouns.

Design is the abstraction and specification of patterns and organs of functionality that have been or will be implemented. Architecture is a degree higher in both abstraction and granularity. Consequentially, architecture is also more topological in nature than design, in that it specifies where major components meet and how they relate to one another. Architecture focuses on the partitioning of major regions of functionality into high level components, where they will physically or virtually reside, what off-the-shelf components may be employed effectively, in general what interfaces each component will expose, what protocols will be employed between them, and what practices and high level patterns may best meet extensibility, maintainability, reliability, durability, scalability, and other non-functional objectives. Design is a detailing of these choices and a more concrete clarification of how functional requirements will be met through the delegation of pieces of that functionality to more granular components and how these smaller components will be organized within the larger ones.

Oftentimes, a portion of architecture is done during the conceptualization of an application, system, or network and may appear in the non-functional sections of requirement documentation. Canonically, design is not specified in requirements, but is rather driven by them.

The process of defining an architecture may involve heuristics, acquired by the architect or architectural team through experience within the domain. As with design, architecture often evolves through a series of iterations, and just as the wisdom of a high level design is often tested when low level design and implementation occurs, the wisdom of an architecture is tested during the specification of a high level design. In both cases, if the wisdom of the specification is called into question during detailing, another iteration of either architecture or design, as the case may be, may become necessary.

In summary, the primary differences between architecture and design are ones of granularity and abstraction, and (consequentially) chronology. (Architecture generally precedes design, although overlap and circular iteration is a common reality).

Examples:

Below the list gives the candidates for being the best ADL to date:

- Primary candidates:
  - ABACUS (UTS),
  - ACME / ADML (CMU/USC),
  - ADML (No longer in development),
  - ByADL (Build Your ADL) - University of L'Aquila,
  - LePUS3 and Class-Z (University of Essex),
  - Rapide (Stanford),
  - Wright (CMU),
  - Unicon (CMU).
- Secondary candidates:
  - Aesop (CMU),
  - MetaH (Honeywell),
  - AADL (SAE) - Architecture Analysis & Design Language,
  - C2 SADL (UCI),
  - SADL (SRI) - System Architecture Description Language.
- Others (unclassified):
  - Lileanna - Library Interconnect Language Extended with Annotated Ada,
  - Dually: Providing Architectural Languages and Tools Interoperability through Model Transformation Technologies,
  - ArchC SystemC-like, focus on instruction sets & memory models,
  - AO-ADL,
  - ArchiMate An example of an ADL for enterprise architecture,
  - DAOP-ADL,
  - DEMO Another example of an enterprise architecture ADL,
  - DiaSpec an approach and tool to generate a distributed framework from a software architecture,
  - SSEP,

    - Unicon,
    - xADL.

Approaches to Architecture

- Academic approach:
    - Focus on analytic evaluation of architectural models.
    - Individual models.
    - Rigorous modeling notations.
    - Powerful analysis techniques.
    - Depth over breadth.
    - Special-purpose solutions.
- Industrial approach:
    - Focus on wide range of development issues.
    - Families of models.
    - Practicality over rigor.
    - Architecture as the big picture in development.
    - Breadth over depth.
    - General-purpose solutions.

EAST-ADL

EAST-ADL is an Architecture Description Language (ADL) for automotive embedded systems, developed in several European research projects. It is designed to complement AUTOSAR with descriptions at higher level of abstractions. Aspects covered by EAST-ADL include vehicle features, functions, requirements, variability, software components, hardware components and communication. Currently, it is maintained by the EAST-ADL Association in cooperation with the European FP7 MAENAD project.

EAST-ADL is a domain-specific language using meta-modeling constructs such as classes, attributes, and relationships. It is based on concepts from UML, SysML and AADL, but adapted for automotive needs and compliance with AUTOSAR. There is an EAST-ADL UML2 profile which is used in UML2 tools for user modeling. The EAST-ADL definition also serves as the specification for implementation in domain-specific tools.

EAST-ADL contains several abstraction levels. The software- and electronics-based functionality of the vehicle are described at different levels of abstraction. The proposed abstraction levels and the contained elements provide a separation of concerns and an implicit style for using the modeling elements. The embedded system is complete on each abstraction level, and parts of the model are linked with various traceability relations. This makes it possible to trace an entity from feature down to components in hardware and software.

EAST-ADL is defined with the development of safety-related embedded control systems as a benchmark. The EAST-ADL scope comprises support for the main phases of software development, from early analysis via functional design to the implementation and back to integration and validation on vehicle level. The main role of EAST-ADL is that of providing an integrated system model. On this basis, several concerns are addressed:

- Documentation, in terms of an integrated system model.
- Communication between engineers, by providing predefined views as well as related information.
- Analysis, through the description of system structure and properties.

Behavioural models for simulation or code generation are supported as references from EAST-ADL functions to external models, such as a subsystem in MATLAB/Simulink.

Organisation of EAST-ADL Meta-model

The EAST-ADL meta-model is organized according to 4 abstraction levels:

- Vehicle level contains modeling elements to represent intended functionality in a solution-independent way.
- Analysis level represents the abstract functional decomposition of the vehicle with the principal internal and external interfaces.
- Design level has the detailed functional definition, a hardware architecture and allocations of functions to hardware.
- Implementation level relies on AUTOSAR elements and does not have EAST-ADL-specific constructs for the core structure.

For all abstraction levels, relevant extension elements for requirements, behavior, variability and dependability are associated to the core structure.

Relation between EAST-ADL and AUTOSAR

Instead of providing modeling entities for the lowest abstraction level, i.e. implementation level, EAST-ADL uses unmodified AUTOSAR entities for this purpose and provides means to link EAST-ADL elements on higher abstraction levels to AUTOSAR elements. Thus, EAST-ADL and AUTOSAR in concert provide means for efficient development and management of the complexity of automotive embedded systems from early analysis right down to implementation. Concepts from model-based development and component-based development reinforce one another. An early, high-level representation of the system can evolve seamlessly into the detailed specifications of the AUTOSAR language. In addition, the EAST-ADL incorporates the following system development concerns:

- Modeling of requirements and verification/validation information.
- Feature modeling and support for software system product lines.
- Modeling of variability of the system design.
- Structural and behavioral modeling of functions and hardware entities in the context of distributed systems.
- Environment, i.e., plant model and adjacent systems.
- Non-functional operational properties such as a definition of function timing and failure modes, supporting system level analysis.

The EAST-ADL metamodel is specified according to the same rules as the AUTOSAR metamodel, which means that the two sets of elements can co-exist in the same model. The dependency is unidirectional from EAST-ADL to AUTOSAR, such that AUTOSAR is independent of EAST-ADL. However, relevant EAST-ADL elements can reference AUTOSAR elements to provide EAST-ADL support for requirements, variability, safety, etc. to the AUTOSAR domain.

A model may thus be defined where AUTOSAR elements represent the software architecture and EAST-ADL elements extend the AUTOSAR model with orthogonal aspects and represents abstract system information through e.g. function and feature models. Such model can be defined in UML, by applying both an EAST-ADL profile and an AUTOSAR profile, or in a domain specific tool based on a merged AUTOSAR and EAST-ADL metamodel.

Modeling Tools and File Format

EAST-ADL tool support is still limited, although a UML profile is available and domain specific tools such as MentorGraphics VSA, MetaCase MetaEdit+ and Systemite SystemWeaver have been tailored for EAST-ADL in the context of research projects and with customers. Papyrus UML, extended within the ATESST project as a concept demonstrator has EAST-ADL support, and MagicDraw, can also provide EAST-ADL palettes, diagrams, etc. In the case of UML, developers also need to have knowledge of UML (classes, stereotypes, arrow types, ..) for modeling with EAST-ADL. Many automotive engineers, in particular mechanical engineers, hardware developers, process experts) do not have this knowledge and prefer other approaches. EATOP is an upcoming initiative to make an Eclipse-based implementation of the EAST-ADL meta-model.

An XML-based exchange format, EAXML, allows tools to exchange EAST-ADL models. The EAXML schema is autogenerated from the EAST-ADL metamodel according to the same principles as the AUTOSAR ARXML schema. Currently, the exchange format is supported by the EAST-ADL prototype of Mentor Graphics VSA, MetaEdit+ and SystemWeaver. For UML tooling, it is possible to exchange models using XMI, subject to the XMI compatibility between tools.

Lifecycle Modeling Language

The Lifecycle Modeling Language (LML) is an open-standard modeling language designed for systems engineering. It supports the full lifecycle: conceptual, utilization, support and retirement stages. Along with the integration of all lifecycle disciplines including, program management, systems and design engineering, verification and validation, deployment and maintenance into one framework. LML was originally designed by the LML steering committee.

The purpose behind the language is to replace predecessor languages, such as UML and SysML that overcomplicate the systems engineering process. LML uses common language to define its modeling elements such as entity, attribute, schedule, cost, and relationship.

LML communicates cost, schedule and performance to all stakeholders in the system lifecycle. LML combines the logical constructs with an ontology to capture information. SysML is mainly constructs and has a limited ontology, while DoDAF MetaModel 2.0 (DM2) only has an ontology. Instead LML simplifies both the constructs and ontology to make them more complete, but still easier to use. There are only 12 primary entity classes. Almost all of the classes relate to each other and themselves with consistent words, i.e., Asset performs Action. Action performed by Asset. SysML uses object oriented design, because it was designed to relate systems thinking to software development. No other discipline in the lifecycle uses object oriented design and analysis extensively. LML captures the entire lifecycle from cradle to grave.

Systems Engineers have identified complexity as a major issue. LML is a new approach to analyzing, planning, specifying, designing, building and maintaining modern systems. LML focuses on these 6 goals:

- To be easy to understand.
- To be easy to extend.

- To support both functional and object oriented approaches within the same design.
- To be a language that can be understood by most system stakeholders, not just Systems Engineers.
- To support systems from cradle to grave.
- To support both evolutionary and revolutionary changes to system plans and designs over the lifetime of a system.

Ontology

Ontologies provide a set of defined terms and relationships between the terms to capture the information that describes the physical, functional, performance, and programmatic aspects of the system. Common ways for describing such ontologies are "Entity", "Relationship", and "Attribute" (ERA). ERA is often used to define database schemas. LML extends the ERA schema with "Attributes on Relationship", a feature that can reduce the number of required "Relationships", in the same way that "Attribute" reduce the number of required "Entities" in ERA. In alignment with the first goal of LML, "Entity", "Relationship", "Attribute", and "Attribute on Relationship" have equivalent English language elements: noun, verb, adjective and adverb.

- Entity (noun): An entity is defined as something that is uniquely identifiable and can exist by itself. There are only 12 parent entities in LML: Action, Artifact, Asset, Characteristic, Connection, Cost, Decision, Input/Output, Location, Risk, Statement and Time. Several child entities have been defined to capture information that stakeholders need. The child entities have the attributes and relationships of the parents plus additional attributes and relationships that make them unique. Child entities include: Conduit (child of Connection), Logical (child of Connection), Measure (child of Characteristic), Orbital (child of Location), Physical (child of Location), Requirement (child of Statement), Resource (child of Asset), and Virtual (child of Location). Every entity has a name or number or description attribute or combination of the three to identify it uniquely. The name is a word or small collection of words providing an overview of information about the entity. The number provides a numerical way to identify the entity. The description provides more detail about that entity.
- Attribute (adjective): The attributes work in the same way an adjective. Entities (the nouns) can have names, numbers, and description attributes. The inherent characteristic or quality of an entity is an attribute. Every attribute has a name that identifies it uniquely within an entity. Attributes names are unique within an entity, but may be used in other entities. The name provides an overview of information about the attribute. The attribute data type specifies the data associated with the attribute.

- Relationship (verb): The relationship works the same way a verb connects nouns or in this case the entities. The relationships enable a simple method to see how [entities] connect. For example, when connecting an action to a statement, LML uses "traced from" as the relationship: an Action is traced from a Statement. The inverse relation of traced from is "traced to." Relationships are defined in both directions and have unique names with the same verb. The standard parent child relationship is decomposed by and its inverse is decomposes. Relationship names are unique across the whole schema.
- Attributes on Relationships (adverb): Classic ERA modeling does not include "attributes on relationships", but is included in LML. In terms of the English language, an "attribute on a relationship" is like an adverb, helping to describe the relationship. Analogous to the way in which attributes relate to entities the "attribute on a relationship" has a name that is unique to its relationship, but need not be unique across other relationships.

3
Systems Development

The process that deals with the design, testing and implementation of a new software program is called systems development. Its focus areas include axiomatic design, IPO model, HIPO model, N2 diagram, etc. The topics elaborated in this chapter will help in gaining a better perspective of the related concepts of systems development.

Systems development is the process of defining, designing, testing, and implementing a new software application or program. It could include the internal development of customized systems, the creation of database systems, or the acquisition of third party developed software. Written standards and procedures must guide all information systems processing functions. The organization's management must define and implement standards and adopt an appropriate system development life cycle methodology governing the process of developing, acquiring, implementing, and maintaining computerized information systems and related technology.

Systems Development Life Cycle

The systems development life cycle (SDLC) is a conceptual model used in project management that describes the stages involved in an information system development project, from an initial feasibility study through maintenance of the completed application. SDLC can apply to technical and non-technical systems. In most use cases, a system is an IT technology such as hardware and software. Project and program managers typically take part in SDLC, along with system and software engineers, development teams and end-users.

Every hardware or software system will go through a development process which can be thought as an iterative process with multiple steps. SDLC is used to give a rigid structure and framework to define the phases and steps involved in the development of a system.

SDLC is also an abbreviation for Synchronous Data Link Control and software development life cycle. Software development life cycle is a very similar process to systems development life cycle, but it focuses exclusively on the development life cycle of software.

Phases

The system development life cycle framework provides a sequence of activities for system designers and developers to follow. It consists of a set of steps or phases in which each phase of the SDLC uses the results of the previous one.

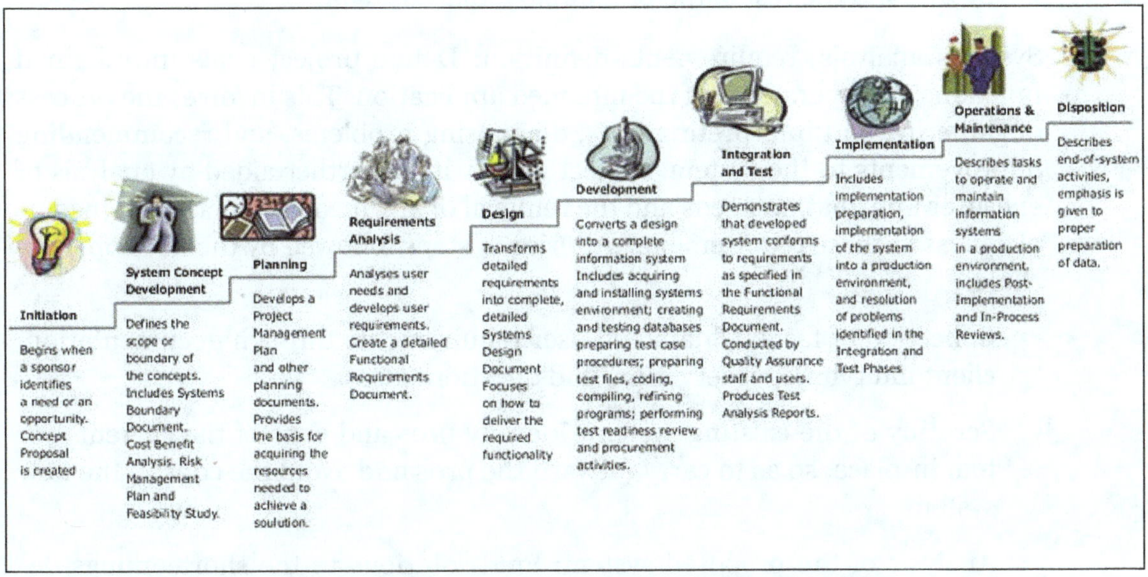

A ten-phase version of the systems development life cycle.

The SDLC adheres to important phases that are essential for developers—such as planning, analysis, design, and implementation. This includes evaluation of the currently used system, information gathering, feasibility studies, and request approval. A number of SDLC models have been created, including waterfall, fountain, spiral, build and fix, rapid prototyping, incremental, synchronize, and stabilize. The oldest of these, and the best known, is the waterfall model, a sequence of stages in which the output of each stage becomes the input for the next. These stages can be characterized and divided up in different ways, including the following:

- Preliminary analysis: Begin with a preliminary analysis, propose alternative solutions, describe costs and benefits, and submit a preliminary plan with recommendations.
    - Conduct the preliminary analysis: Discover the organization's objectives and the nature and scope of the problem under study. Even if a problem refers only to a small segment of the organization itself, find out what the objectives of the organization itself are. Then see how the problem being studied fits in with them.

    - Propose alternative solutions: After digging into the organization's objectives and specific problems, several solutions may have been discovered. However, alternate proposals may still come from interviewing employees, clients, suppliers, and/or consultants. Insight may also be gained by researching what competitors are doing.
    - Cost benefit analysis: Analyze and describe the costs and benefits of implementing the proposed changes. In the end, the ultimate decision on whether to leave the system as is, improve it, or develop a new system will be guided by this and the rest of the preliminary analysis data.
- Systems analysis, requirements definition: Define project goals into defined functions and operations of the intended application. This involves the process of gathering and interpreting facts, diagnosing problems, and recommending improvements to the system. Project goals will be further aided by analysis of end-user information needs and the removal of any inconsistencies and incompleteness in these requirements. A series of steps followed by the developer include:
    - Collection of facts: Obtain end user requirements through documentation, client interviews, observation, and questionnaires.
    - Scrutiny of the existing system: Identify pros and cons of the current system in-place, so as to carry forward the pros and avoid the cons in the new system.
    - Analysis of the proposed system: Find solutions to the shortcomings described in step two and prepare the specifications using any specific user proposals.
- Systems design: At this step, desired features and operations are described in detail, including screen layouts, business rules, process diagrams, pseudocode, and other documentation.
- Development: The real code is written here.
- Integration and testing: All the modules are brought together into a special testing environment, then checked for errors, bugs, and interoperability.
- Acceptance, installation, deployment: This is the final stage of initial development, where the software is put into production and runs actual business.
- Maintenance: During the maintenance stage of the SDLC, the system is assessed/evaluated to ensure it does not become obsolete. This is also where changes are made to initial software.
- Evaluation: Some companies do not view this as an official stage of the SDLC, while others consider it to be an extension of the maintenance stage, and may

be referred to in some circles as post-implementation review. This is where the system that was developed, as well as the entire process, is evaluated. Some of the questions that need to be answered include if the newly implemented system meets the initial business requirements and objectives, if the system is reliable and fault-tolerant, and if it functions according to the approved functional requirements. In addition to evaluating the software that was released, it is important to assess the effectiveness of the development process. If there are any aspects of the entire process (or certain stages) that management is not satisfied with, this is the time to improve.

- Disposal: In this phase, plans are developed for discontinuing the use of system information, hardware, and software and making the transition to a new system. The purpose here is to properly move, archive, discard, or destroy information, hardware, and software that is being replaced, in a manner that prevents any possibility of unauthorized disclosure of sensitive data. The disposal activities ensure proper migration to a new system. Particular emphasis is given to proper preservation and archiving of data processed by the previous system. All of this should be done in accordance with the organization's security requirements.

These stages of the systems development life cycle are divided in ten steps, from definition to creation and modification of IT work products. Not every project will require that the phases be sequentially executed. However, the phases are interdependent. Depending upon the size and complexity of the project, phases may be combined or may overlap.

System Investigation

First the IT system proposal is investigated. During this step, consider all current priorities that would be affected and how they should be handled. Before any system planning is done, a feasibility study should be conducted to determine if creating a new or improved system is a viable solution. This will help to determine the costs, benefits, resource requirements, and specific user needs required for completion. The development process can only continue once management approves of the recommendations from the feasibility study.

The following represent different components of the feasibility study:

- Operational feasibility,
- Financial feasibility,
- Technical feasibility,
- Human factors feasibility,
- Legal/Political feasibility.

Analysis

The goal of analysis is to determine where the problem is, in an attempt to fix the system. This step involves breaking down the system in different pieces to analyze the situation, analyzing project goals, breaking down what needs to be created, and attempting to engage users so that definite requirements can be defined.

Design

In systems design, the design functions and operations are described in detail, including screen layouts, business rules, process diagrams, and other documentation. The output of this stage will describe the new system as a collection of modules or subsystems.

The design stage takes as its initial input the requirements identified in the approved requirements document. For each requirement, a set of one or more design elements will be produced as a result of interviews, workshops, and/or prototype efforts.

Design elements describe the desired system features in detail, and they generally include functional hierarchy diagrams, screen layout diagrams, tables of business rules, business process diagrams, pseudo-code, and a complete entity-relationship diagram with a full data dictionary. These design elements are intended to describe the system in sufficient detail, such that skilled developers and engineers may develop and deliver the system with minimal additional input design.

Environments

Environments are controlled areas where systems developers can build, distribute, install, configure, test, and execute systems that move through the SDLC. Each environment is aligned with different areas of the SDLC and is intended to have specific purposes. Examples of such environments include the:

- Development environment, where developers can work independently of each other before trying to merge their work with the work of others.
- Common build environment, where merged work can be built, together, as a combined system.
- Systems integration testing environment, where basic testing of a system's integration points to other upstream or downstream systems can be tested.
- User acceptance testing environment, where business stakeholders can test against their original business requirements.
- Production environment, where systems finally get deployed for final use by their intended end users.

Testing

The code is tested at various levels in software testing. Unit, system, and user acceptance testings are often performed. This is a grey area as many different opinions exist as to what the stages of testing are and how much, if any iteration occurs. Iteration is not generally part of the waterfall model, but the means to rectify defects and validate fixes prior to deployment is incorporated into this phase.

The following are types of testing that may be relevant, depending on the type of system under development:

- Defect testing the failed scenarios, including:
    - Path testing,
    - Data set testing,
    - Unit testing,
    - System testing,
    - Integration testing,
    - Black-box testing,
    - White-box testing,
    - Regression testing,
    - Automation testing,
    - User acceptance testing,
    - Software performance testing.

Training and Transition

Once a system has been stabilized through adequate testing, the SDLC ensures that proper training on the system is performed or documented before transitioning the system to its support staff and end users. Training usually covers operational training for those people who will be responsible for supporting the system as well as training for those end users who will be using the system after its delivery to a production operating environment.

After training has been successfully completed, systems engineers and developers transition the system to its final production environment, where it is intended to be used by its end users and supported by its support and operations staff.

Operations and Maintenance

The deployment of the system includes changes and enhancements before the decommissioning or sunset of the system. Maintaining the system is an important aspect of SDLC. As key personnel change positions in the organization, new changes will be implemented. There are two approaches to system development: the traditional approach (structured) and object oriented. Information engineering includes the traditional system approach, which is also called the structured analysis and design technique. The object oriented approach views information system as a collection of objects that are integrated with each other to make a full and complete information system.

Evaluation

The final phase of the SDLC is to measure the effectiveness of the system and evaluate potential enhancements.

Systems Analysis and Design

The systems analysis and design (SAD) is the process of developing information systems (IS) that effectively use hardware, software, data, processes, and people to support the company's businesses objectives. System analysis and design can be considered the meta-development activity, which serves to set the stage and bound the problem. SAD can be leveraged to set the correct balance among competing high-level requirements in the functional and non-functional analysis domains. System analysis and design interacts strongly with distributed enterprise architecture, enterprise I.T. Architecture, and business architecture, and relies heavily on concepts such as partitioning, interfaces, personae and roles, and deployment/operational modeling to arrive at a high-level system description. This high level description is then further broken down into the components and modules which can be analyzed, designed, and constructed separately and integrated to accomplish the business goal. SDLC and SAD are cornerstones of full life cycle product and system planning.

Object-oriented Analysis

Object-oriented analysis (OOA) is the process of analyzing a task (also known as a problem domain), to develop a conceptual model that can then be used to complete the task. A typical OOA model would describe computer software that could be used to satisfy a set of customer-defined requirements. During the analysis phase of problem-solving, a programmer might consider a written requirements statement, a formal vision document, or interviews with stakeholders or other interested parties. The task to be addressed might be divided into several subtasks (or domains), each representing a different business, technological, or other areas of interest. Each subtask would be analyzed separately. Implementation constraints, (e.g., concurrency, distribution, persistence, or how the system is to be built) are not considered during the analysis phase; rather, they are addressed during object-oriented design (OOD).

The conceptual model that results from OOA will typically consist of a set of use cases, one or more UML class diagrams, and a number of interaction diagrams. It may also include some kind of user interface mock-up.

The input for object-oriented design is provided by the output of object-oriented analysis. Realize that an output artifact does not need to be completely developed to serve as input of object-oriented design; analysis and design may occur in parallel, and in practice the results of one activity can feed the other in a short feedback cycle through an iterative process. Both analysis and design can be performed incrementally, and the artifacts can be continuously grown instead of completely developed in one shot. Some typical (but common to all types of design analysis) input artifacts for object-oriented:

- Conceptual model: Conceptual model is the result of object-oriented analysis, it captures concepts in the problem domain. The conceptual model is explicitly chosen to be independent of implementation details, such as concurrency or data storage.
- Use case: Use case is a description of sequences of events that, taken together, lead to a system doing something useful. Each use case provides one or more scenarios that convey how the system should interact with the users called actors to achieve a specific business goal or function. Use case actors may be end users or other systems. In many circumstances use cases are further elaborated into use case diagrams. Use case diagrams are used to identify the actor (users or other systems) and the processes they perform.
- System Sequence Diagram: System Sequence diagram (SSD) is a picture that shows, for a particular scenario of a use case, the events that external actors generate, their order, and possible inter-system events.
- User interface documentations (if applicable): Document that shows and describes the look and feel of the end product's user interface. It is not mandatory to have this, but it helps to visualize the end-product and therefore helps the designer.
- Relational data model (if applicable): A data model is an abstract model that describes how data is represented and used. If an object database is not used, the relational data model should usually be created before the design, since the strategy chosen for object-relational mapping is an output of the OO design process. However, it is possible to develop the relational data model and the object-oriented design artifacts in parallel, and the growth of an artifact can stimulate the refinement of other artifacts.

Life Cycle

Management and Control

The SDLC phases serve as a programmatic guide to project activity and provide a flexible but consistent way to conduct projects to a depth matching the scope of the project. Each of the SDLC phase objectives are described here with key deliverables, a description of recommended tasks, and a summary of related control objectives for effective management. It is critical for the project manager to establish and monitor control objectives during each SDLC phase while executing projects. Control objectives help to provide a clear statement of the desired result or purpose and should be used throughout the entire SDLC process. Control objectives can be grouped into major categories (domains), and relate to the SDLC phases as shown in the figure.

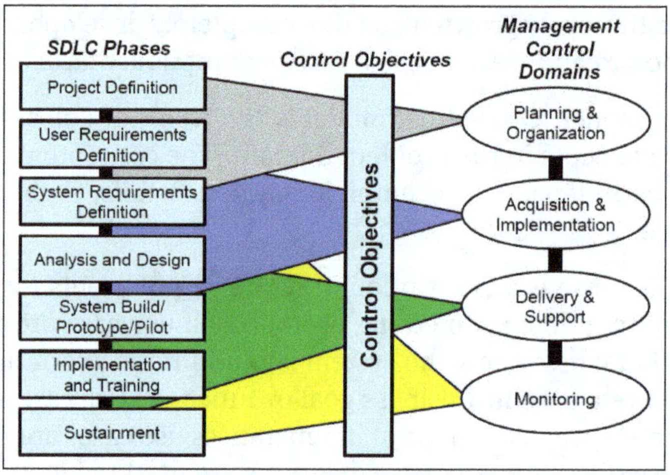

SPIU phases related to management controls.

To manage and control any SDLC initiative, each project will be required to establish some degree of a work breakdown structure (WBS) to capture and schedule the work necessary to complete the project. The WBS and all programmatic material should be kept in the "project description" section of the project notebook. The WBS format is mostly left to the project manager to establish in a way that best describes the project work.

There are some key areas that must be defined in the WBS as part of the SDLC policy. The following diagram describes three key areas that will be addressed in the WBS in a manner established by the project manager. The diagram shows coverage spans numerous phases of the SDLC but the associated MCD has a subset of primary mappings to the SDLC phases. For example, Analysis and Design is primarily performed as part of the Acquisition and Implementation Domain and System Build and Prototype is primarily performed as part of delivery and support.

Work Breakdown Structured Organization

The upper section of the work breakdown structure (WBS) should identify the major phases and milestones of the project in a summary fashion. In addition, the upper section should provide an overview of the full scope and timeline of the project and will be part of the initial project description effort leading to project approval. The middle section of the WBS is based on the seven systems development life cycle phases as a guide for WBS task development. The WBS elements should consist of milestones and "tasks" as opposed to "activities" and have a definitive period (usually two weeks or more). Each task must have a measurable output (e.x. document, decision, or analysis). A WBS task may rely on one or more activities (e.g. software engineering, systems engineering) and may require close coordination with other tasks, either internal or external to the project. Any part of the project needing support from contractors should have a statement of work (SOW) written to include the appropriate tasks from the SDLC phases. The development of a SOW does not occur during a specific phase of SDLC but is developed to include the work from the SDLC process that may be conducted by external resources such as contractors.

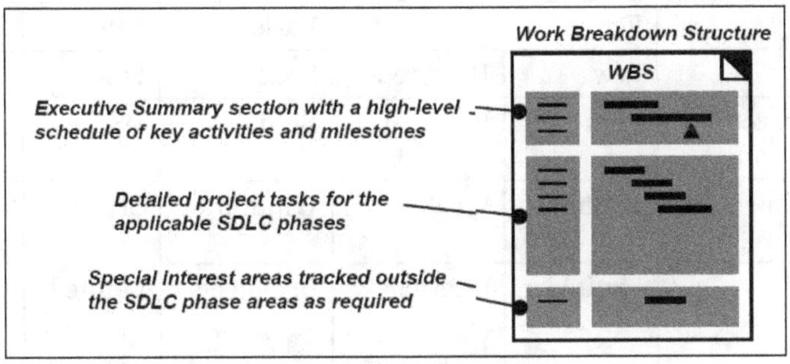

Work breakdown structure.

Baselines

Baselines are an important part of the systems development life cycle. These baselines are established after four of the five phases of the SDLC and are critical to the iterative nature of the model. Each baseline is considered as a milestone in the SDLC.

- Functional baseline: Established after the conceptual design phase.
- Allocated baseline: Established after the preliminary design phase.
- Product baseline: Established after the detail design and development phase.
- Updated product baseline: Established after the production construction phase.

Complementary Methodologies

Complementary software development methods to systems development life cycle are:

- Software prototyping.
- Joint applications development (JAD).
- Rapid application development (RAD).
- Extreme programming (XP).

- Open-source development.
- End-user development.
- Object-oriented programming.

Table: Comparison of Methodology Approaches.

	SDLC	RAD	Open source	Objects	JAD	Prototyping	End User
Control	Formal	MIS	Weak	Standards	Joint	User	User
Time frame	Long	Short	Medium	Any	Medium	Short	Short –
Users	Many	Few	Few	Varies	Few	One or two	One
MIS staff	Many	Few	Hundreds	Split	Few	One or two	None
Transaction/ DSS	Transaction	Both	Both	Both	DSS	DSS	DSS
Interface	Minimal	Minimal	Weak	Windows	Crucial	Crucial	Crucial
Documentation and training	Vital	Limited	Internal	In Objects	Limited	Weak	None
Integrity and security	Vital	Vital	Unknown	In Objects	Limited	Weak	Weak
Reusability	Limited	Some	Maybe	Vital	Limited	Weak	None

Strengths and Weaknesses

Few people in the modern computing world would use a strict waterfall model for their SDLC as many modern methodologies have superseded this thinking. Some will argue that the SDLC no longer applies to models like Agile computing, but it is still a term widely in use in technology circles. The SDLC practice has advantages in traditional models of systems development that lends itself more to a structured environment. The disadvantages to using the SDLC methodology is when there is need for iterative development or (i.e. web development or e-commerce) where stakeholders need to review on a regular basis the software being designed.

Table: Strength and Weaknesses of SDLC.

Strengths	Weaknesses
Control	Increased development time
Monitor large projects	Increased development cost
Detailed steps	Systems must be defined up front
Evaluate costs and completion targets	Rigidity

Documentation	Hard to estimate costs, project overruns
Well defined user input	User input is sometimes limited
Ease of maintenance	Little parallelism
Development and design standards	Automation of documentation and standards is limited
Tolerates changes in MIS of staffing	Does not tolerate changes in requirements
	Projects canned early on the result in little or no value

An alternative to the SDLC is rapid application development, which combines prototyping, joint application development and implementation of CASE tools. The advantages of RAD are speed, reduced development cost, and active user involvement in the development process.

System Lifecycle

The system lifecycle in systems engineering is a view of a system or proposed system that addresses all phases of its existence to include system conception, design and development, production and/or construction, distribution, operation, maintenance and support, retirement, phase-out and disposal.

Conceptual Design

The conceptual design stage is the stage where an identified need is examined, requirements for potential solutions are defined, potential solutions are evaluated and a system specification is developed. The system specification represents the technical requirements that will provide overall guidance for system design. Because this document determines all future development, the stage cannot be completed until a conceptual design review has determined that the system specification properly addresses the motivating need.

Key steps within the conceptual design stage include:

- Need identification,
- Feasibility analysis,
- System requirements analysis,
- System specification,
- Conceptual design review.

Preliminary System Design

During this stage of the system lifecycle, subsystems that perform the desired system functions are designed and specified in compliance with the system specification. Interfaces between subsystems are defined, as well as overall test and evaluation requirements. At the completion of this stage, a development specification is produced that is sufficient to perform detailed design and development.

Key steps within the preliminary design stage include:

- Functional analysis,
- Requirements allocation,
- Detailed trade-off studies,
- Synthesis of system options,
- Preliminary design of engineering models,
- Development specification,
- Preliminary design review.

For example, as the system analyst of Viti Bank, you have been tasked to examine the current information system. Viti Bank is a fast growing bank in Fiji. Customers in remote rural areas are finding difficulty to access the bank services. It takes them days or even weeks to travel to a location to access the bank services. With the vision of meeting the customer's needs, the bank has requested your services to examine the current system and to come up with solutions or recommendations of how the current system can be provided to meet its needs.

Detail Design and Development

This stage includes the development of detailed designs that brings initial design work into a completed form of specifications. This work includes the specification of interfaces between the system and its intended environment and a comprehensive evaluation of the systems logistical, maintenance and support requirements. The detail design and development is responsible for producing the product, process and material specifications and may result in substantial changes to the development specification.

Key steps within the detail design and development stage include:

- Detailed design,
- Detailed synthesis,
- Development of engineering and prototype models,
- Revision of development specification,
- Product, process and material specification,
- Critical design review.

Production and Construction

During the production and/or construction stage the product is built or assembled in accordance with the requirements specified in the product, process and material specifications and is deployed and tested within the operational target environment. System assessments are conducted in order to correct deficiencies and adapt the system for continued improvement.

Key steps within the product construction stage include:

- Production and construction of system components.
- Acceptance testing.
- System distribution and operation.
- Operational testing and evaluation.
- System assessment.

Utilization and Support

Once fully deployed, the system is used for its intended operational role and maintained within its operational environment.

Key steps within the utilization and support stage include:

- System operation in the user environment.
- Change management.
- System modifications for improvement.
- System assessment.

Phase-out and Disposal

Effectiveness and efficiency of the system must be continuously evaluated to determine when the product has met its maximum effective lifecycle. Considerations include: Continued existence of operational need, matching between operational requirements and system performance, feasibility of system phase-out versus maintenance, and availability of alternative systems.

Advantages and Disadvantages of SDLC

Benefits of abiding by a clearly defined SDLC model include:

- Having a clear view of an entire project, workers involved, estimated costs and timelines.

- Gives project managers a projected base cost of the project.
- Goals and standards are clearly defined.
- Developers can move back a step if something does not go as expected.

Disadvantages, however, can include:

- Due to assumptions made at the beginning of a project, if an unexpected circumstance complicates the development of a system, then it may stockpile into more complications down the road. As an example, if newly installed hardware does not work correctly, then it may increase the time a system is in development, increasing the cost.
- Some methods are not flexible.
- It can be complicated to estimate the overall cost at the beginning of a project.
- Testing at the end of development may slow down some development teams.

Structured Analysis and Design Technique

Structured analysis and design technique (SADT) is a systems engineering and software engineering methodology for describing systems as a hierarchy of functions. SADT is a structured analysis modelling language, which uses two types of diagrams: activity models and data models. It was developed in the late 1960s by Douglas T. Ross, and was formalized and published as IDEF0 in 1981.

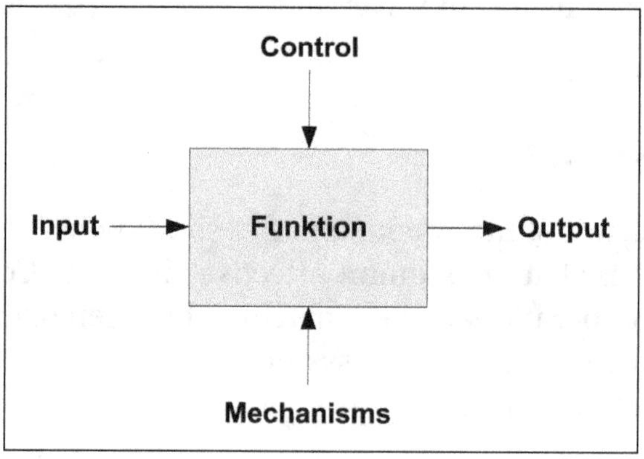

SADT basis element.

Structured analysis and design technique (SADT) is a diagrammatic notation designed specifically to help people describe and understand systems. It offers building blocks to represent entities and activities, and a variety of arrows to relate boxes. These boxes and arrows have an associated informal semantics. SADT can be used as a functional analysis tool of a given process, using successive levels of details. The SADT method not only allows one to define user needs for IT developments, which is often used in the industrial Information Systems, but also to explain and present an activity's manufacturing processes and procedures.

Top-down Approach

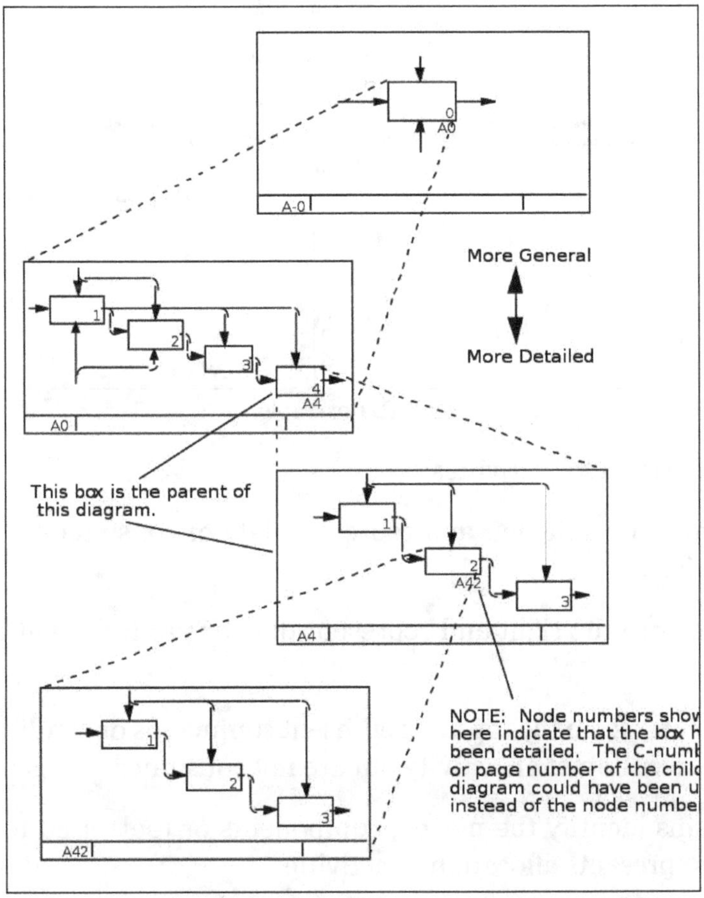

Top down decomposition structure.

The structured analysis and design technique uses a decomposition with the top-down approach. This decomposition is conducted only in the physical domain from an axiomatic design viewpoint.

Diagrams

SADT uses two types of diagrams: activity models and data models. It uses arrows to build these diagrams. The SADT's representation is the following:

- A main box where the name of the process or the action is specified.
- On the left-hand side of this box, incoming arrows: Inputs of the action.

- On the upper part, the incoming arrows: data necessary for the action.
- On the bottom of the box, incoming arrows: means used for the action.
- On the right-hand side of the box, outgoing arrows: outputs of the action.

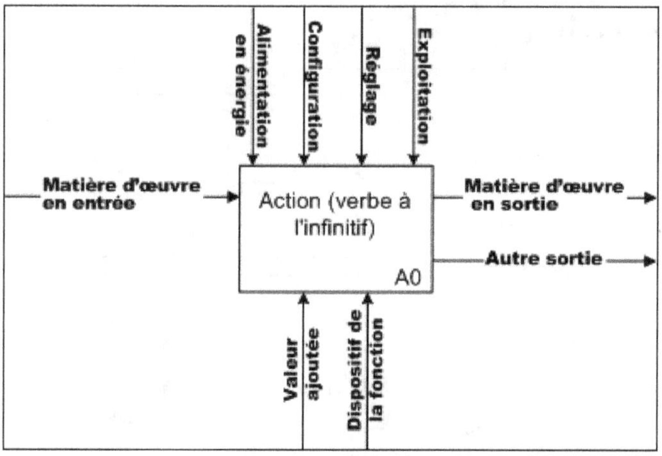

An SADT example.

The semantics of arrows for activities:

- Inputs enter from the left and represent data or consumables that are needed by the activity.
- Outputs exit to the right and represent data or products that are produced by the activity.
- Controls enter from the top and represent commands or conditions which influence the execution of an activity but are not consumed.
- Mechanisms identify the means, components or tools used to accomplish the activity. Represents allocation of activities.

The semantics of arrows for data:

- Inputs are activities that produce the data.
- Outputs consume the data.
- Controls influence the internal state of the data.

Roles

According to Mylopoulos in the software development process multiple roles can or should be distinguished:

- Author or developer of the SADT models.

- Commenters, who review the author's work.
- Readers or users of the SADT models.
- Experts, who can advise the authors.
- Technical committee or reviewers of the SADT models in detail.
- Project librarian, who govern the project documentation.
- Project manager, who governs the system analysis and design.
- Monitor or chief analyst to assists SADT developers and users.
- Instructor to train SADT developers and users.

Usage

SADT is used as diagrammatic notation in conceptual design of software engineering and systems engineering to sketch applications, for more detailed structured analysis, for requirements definition, and structured design.

Axiomatic Design

Axiomatic design is a systems design methodology using matrix methods to systematically analyze the transformation of customer needs into functional requirements, design parameters, and process variables. Specifically, a set of functional requirements(-FRs) are related to a set of design parameters (DPs) by a Design Matrix A:

$$\begin{bmatrix} FR_1 \\ FR_2 \end{bmatrix} = \begin{bmatrix} A_{11} & A_{12} \\ A_{21} & A_{22} \end{bmatrix} \begin{bmatrix} DP_1 \\ DP_2 \end{bmatrix}$$

The method gets its name from its use of design principles or design Axioms (i.e., given without proof) governing the analysis and decision making process in developing high quality product or system designs. The two axioms used in Axiomatic Design (AD) are:

- Axiom 1: The Independence Axiom. Maintain the independence of the functional requirements (FRs).
- Axiom 2: The Information Axiom. Minimize the information content of the design.

Axiomatic design is considered to be a design method that addresses fundamental issues in Taguchi methods.

Coupling is the term Axiomatic Design uses to describe a lack of independence between the FRs of the system as determined by the DPs. i.e., if varying one DP has a resulting significant impact on two separate FRs, it is said the FRs are coupled. Axiomatic Design introduces matrix analysis of the Design Matrix to both assess and mitigate the effects of coupling.

Axiom 2, the Information Axiom, provides a metric of the probability that a specific DP will deliver the functional performance required to satisfy the FR. The metric is normalized to be summed up for the entire system being modeled. Systems with less functional performance risk (minimal information content) are preferred over alternative systems with higher information content.

The methodology was developed by Dr. Suh Nam Pyo at MIT, Department of Mechanical Engineering since the 1990s. A series of academic conferences have been held to present current developments of the methodology.

IPO Model

The input–process–output (IPO) model, or input-process-output pattern, is a widely used approach in systems analysis and software engineering for describing the structure of an information processing program or other process. Many introductory programming and systems analysis texts introduce this as the most basic structure for describing a process.

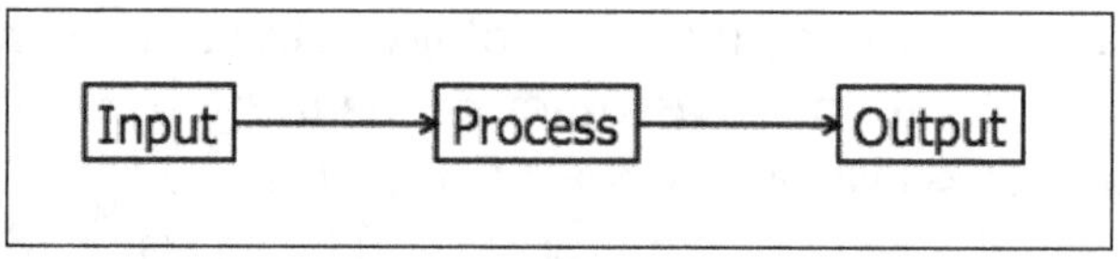

The input–process–output model.

A computer program is useful for another sort of process using the input-process-output model receives inputs from a user or other source, does some computations on the inputs, and returns the results of the computations. In essence the system separates itself from the environment, thus defining both inputs and outputs as one united mechanism. The system would divide the work into three categories:

- A requirement from the environment (input).
- A computation based on the requirement (process).
- A provision for the environment (output).

In other words, such inputs may be materials, human resources, money or information, transformed into outputs, such as consumables, services, new information or money.

As a consequence, an Input-Process-Output system becomes very vulnerable to misinterpretation. This is because, theoretically, it contains all the data, in regards to the environment outside the system. Yet, in practice, the environment contains a significant variety of objects that a system is unable to comprehend, as it exists outside the system's control. As a result, it is very important to understand where the boundary lies between the system and the environment, which is beyond the system's understanding. Various analysts often set their own boundaries, favoring their point of view, thus creating much confusion.

Systems at Work

The views differ, in regards to systems thinking. One of such definitions would outline the Input-process-output system, as a structure, would be:

> "Systems thinking is the art and science of making reliable inferences about behaviour by developing an increasingly deep understanding of the understanding of the underlying structure"

Alternatively, it was also suggested that systems are not 'holistic' in the sense of bonding with remote objects (for example: trying to connect a crab, ozone layer and capital life cycle together).

Types of Systems

There are five major categories that are the most cited in information systems literature:

- Natural systems: A system which has not been created as a result of human interference. Examples of such would be the solar system as well as the human body, evolving into its current form.
- Designed physical systems: A system which has been created as a result of human interference, and is physically identifiable. Examples of such would be various computing machines, created by human mind for some specific purpose.
- Designed abstract systems: A system which has been created as a result of human interference, and is not physically identifiable. Examples of such would be mathematical and philosophical systems, which have been created by human minds, for some specific purpose.
    - There are also some social systems, which allow humans to collectively achieve a specific purpose.
- Social systems: A system created by humans, and derived from intangible purposes. For example: a family, that is a hierarchy of human relationships, which in essence create the boundary between natural and human systems.
- Human activity systems: An organisation with hierarchy, created by humans for a specific purpose. For example: a company, which organises humans

together to collaborate and achieve a specific purpose. The result of this system is physically identifiable. There are, however, some significant links between with previous types. It is clear that the idea of human activity system (HAS), would consist of a variety of smaller social system, with its unique development and organisation. Moreover, arguably HASes can include designed systems - computers and machinery. Majority of previous systems would overlap.

System Characteristics

There are several key characteristics, when it comes to the fundamental behaviour of any system:

- Systems can be classified as open or closed:
    - Those that interact with their environment, in form of money, data, energy or exchange materials, are generally understood as open. Openness of the system can vary significantly. This is because, a system would be classified as open, if it receives even a single input from the environment, yet a system that merely interacts with the environment, would be classified as open as well. The more open the system is, the more complex it normally would be, due to lower predictability of its components.
    - Those that have no interactions with the environment at all are closed. In practice, however, a completely closed system is merely liveable, due to loss of practical usage of the output. As a result, most of the systems would be open or open to a certain extent.
- Systems can be classified as deterministic or stochastic:
    - Well-defined and clearly structured system in terms of behavioural patterns becomes predictable, thus becoming deterministic. In other words it would only use empirical data. For example: mathematics or physics are set around specific laws, which make the results of calculation predictable. Deterministic systems would have simplistic interactions between inner components.
    - More complex, and often more open systems, would have relatively lower extent of predictability, due to absence of clearly structured behavioural patterns. Analysing such system, is therefore much harder. Such systems would be stochastic, or probabilistic, this is because of the stochastic nature of human beings whilst performing various activities. Having said that, designed systems would still be considered as deterministic, due to a rigid structure of rules incorporated into the design.
- Systems can be classified as static or dynamic:
    - Most systems would be known as dynamic, because of the constant evolution in computing power, yet some systems could find it hard to balance

between being created and ceasing to exist. An example of such could be a printed map, which is not evolving, in contrast to a dynamic map, provided from constantly updating developers.

- Systems can be classified as self-regulating or non-self-regulating:
    - The greater the extent of self-control of systems activity is, the greater is the liveability of the final system is. It is vital for any system to be able to control its activities in order to remain stable.

Real Life Applications

- Corporate business: A manufacturing processes that take raw materials as inputs, applies a manufacturing process, and produces manufactured goods as output. The usage of such systems could help to create stronger human organisations, in terms of company operations in each and every department of the firm, no matter the size, which IPOs can also restructure existing static and non-self-regulating systems, which in real world would be used in form of outsourcing the product fulfilment, due to inefficiency of current fulfilment.
- Programming:
    - The majority of existing programs for coding, such as Java, Python, C++, would be based upon a deterministic IPO model, with clear inputs coming from the coder, converting into outputs, such as applications.
    - A batch transaction processing system, which accepts large volumes of homogeneous transactions, processes it (possibly updating a database), and produces output such as reports or computations.
    - An interactive computer program, which accepts simple requests from a user and responds to them after some processing and/or database accesses.
- Scientific:
    - A calculator, which uses inputs, provided by the operator, and processes them into outputs to be used by the operator.
    - A thermostat, which senses the temperature (input), decides on an action (heat on/off), and executes the action (output).

Function Model

In systems engineering, software engineering, and computer science, a function model or functional model is a structured representation of the functions (activities, actions, processes, operations) within the modeled system or subject area.

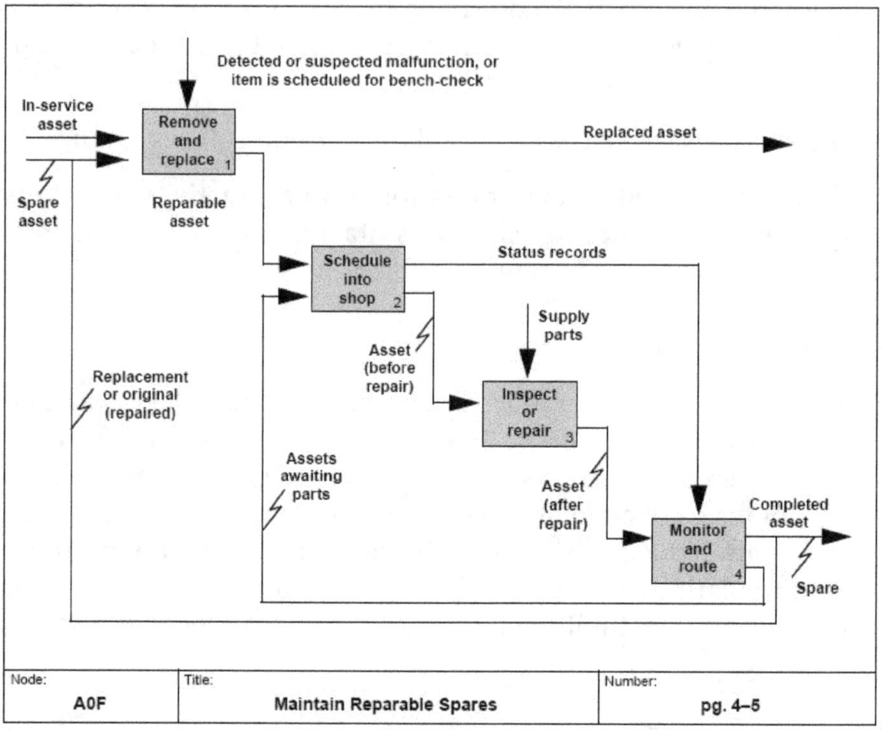

Example of a function model of the process of "Maintain Reparable Spares" in IDEF0 notation.

A function model, similar with the activity model or process model, is a graphical representation of an enterprise's function within a defined scope. The purposes of the function model are to describe the functions and processes, assist with discovery of information needs, help identify opportunities, and establish a basis for determining product and service costs.

Functional Modeling Methods

Functional Flow Block Diagram

A functional flow block diagram (FFBD) is a multi-tier, time-sequenced, step-by-step flow diagram of a system's functional flow. The term "functional" in this context is different from its use in functional programming or in mathematics, where pairing "functional" with "flow" would be ambiguous. Here, "functional flow" pertains to the sequencing of operations, with "flow" arrows expressing dependence on the success of prior operations. FFBDs may also express input and output data dependencies between functional blocks, as shown in figures below, but FFBDs primarily focus on sequencing.

The FFBD notation was developed in the 1950s, and is widely used in classical systems engineering. FFBDs are one of the classic business process modeling methodologies, along with flow charts, data flow diagrams, control flow diagrams, Gantt charts, PERT diagrams, and IDEF.

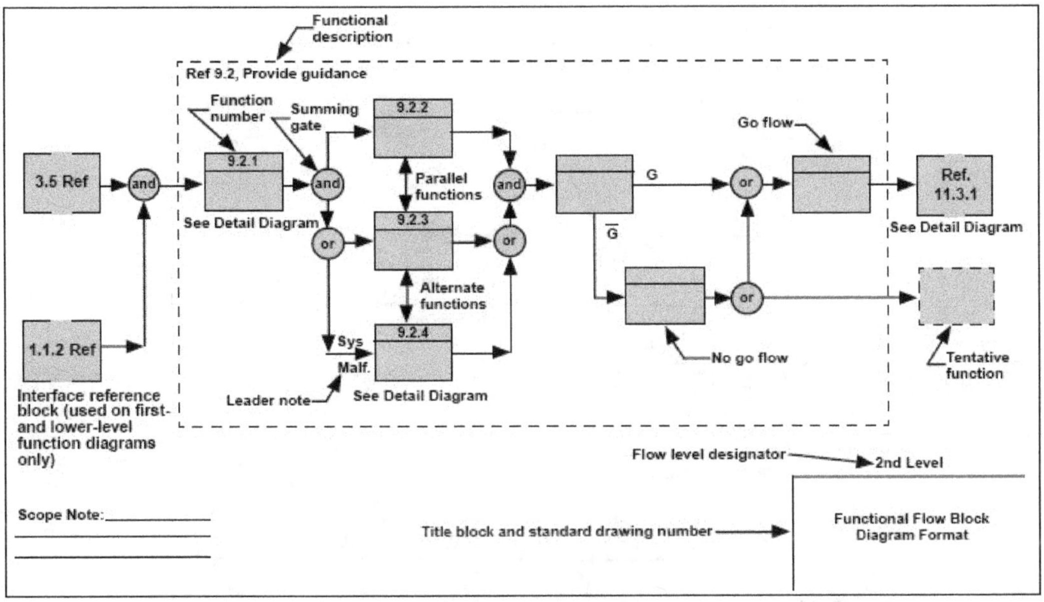

Functional Flow Block Diagram Format.

FFBDs are also referred to as Functional Flow Diagrams, functional block diagrams, and functional flows.

Development of Functional Flow Block Diagrams

FFBDs can be developed in a series of levels. FFBDs show the same tasks identified through functional decomposition and display them in their logical, sequential relationship. For example, the entire flight mission of a spacecraft can be defined in a top level FFBD. Each block in the first level diagram can then be expanded to a series of functions, as shown in the second level diagram for "perform mission operations." The diagram shows both input (transfer to operational orbit) and output (transfer to space transportation system orbit), thus initiating the interface identification and control process. Each block in the second level diagram can be progressively developed into a series of functions.

These diagrams are used both to develop requirements and to identify profitable trade studies. For example, does the spacecraft antenna acquire the tracking and data relay satellite (TDRS) only when the payload data are to be transmitted, or does it track TDRS continually to allow for the reception of emergency commands or transmission of emergency data? The FFBD also incorporates alternate and contingency operations, which improve the probability of mission success. The flow diagram provides an understanding of total operation of the system, serves as a basis for development of operational and contingency procedures, and pinpoints areas where changes in operational procedures could simplify the overall system operation. In certain cases, alternate FFBDs may be used to represent various means of satisfying a particular function until data are acquired, which permits selection among the alternatives.

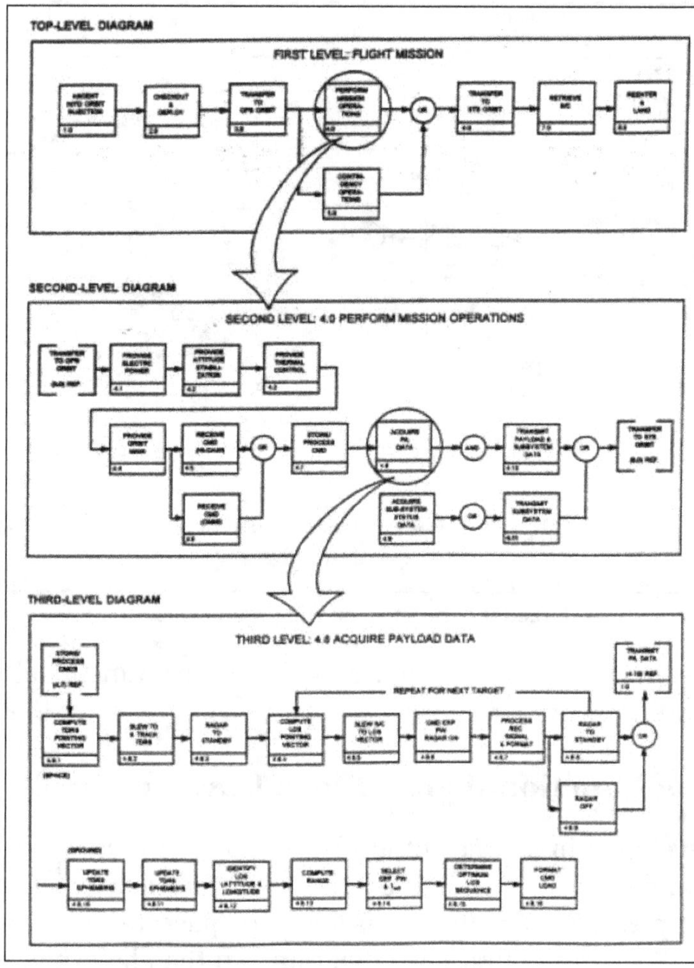

Development of Functional Flow Block Diagrams.

Building Blocks

Key Attributes

An overview of the key FFBD attributes:

- Function block: Each function on an FFBD should be separate and be represented by single box (solid line). Each function needs to stand for definite, finite, discrete action to be accomplished by system elements.
- Function numbering: Each level should have a consistent number scheme and provide information concerning function origin. These numbers establish identification and relationships that will carry through all Functional Analysis and Allocation activities and facilitate traceability from lower to top levels.
- Functional reference: Each diagram should contain a reference to other functional diagrams by using a functional reference (box in brackets).

- Flow connection: Lines connecting functions should only indicate function flow and not a lapse in time or intermediate activity.
- Flow direction: Diagrams should be laid out so that the flow direction is generally from left to right. Arrows are often used to indicate functional flows.
- Summing gates: A circle is used to denote a summing gate and is used when AND/OR is present. AND is used to indicate parallel functions and all conditions must be satisfied to proceed. OR is used to indicate that alternative paths can be satisfied to proceed.
- GO and NO-GO paths: "G" and "bar G" are used to denote "go" and "no-go" conditions. These symbols are placed adjacent to lines leaving a particular function to indicate alternative paths.

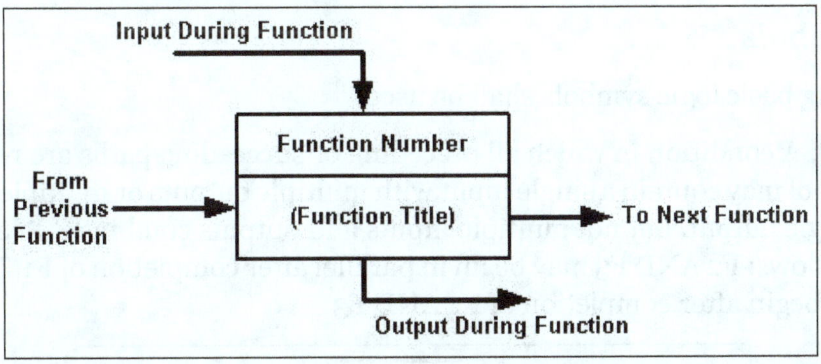

Graphical explanation of a "function block" used in these diagrams. Flow is from left to right.

Function Symbolism

A function shall be represented by a rectangle containing the title of the function (an action verb followed by a noun phrase) and its unique decimal delimited number. A horizontal line shall separate this number and the title. The figure also depicts how to represent a reference function, which provides context within a specific FFBD.

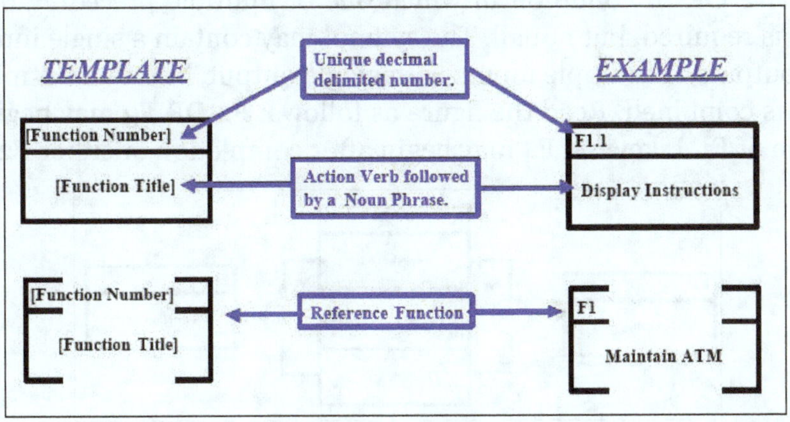

Function Symbol.

Directed Lines

A line with a single arrowhead shall depict functional flow from left to right.

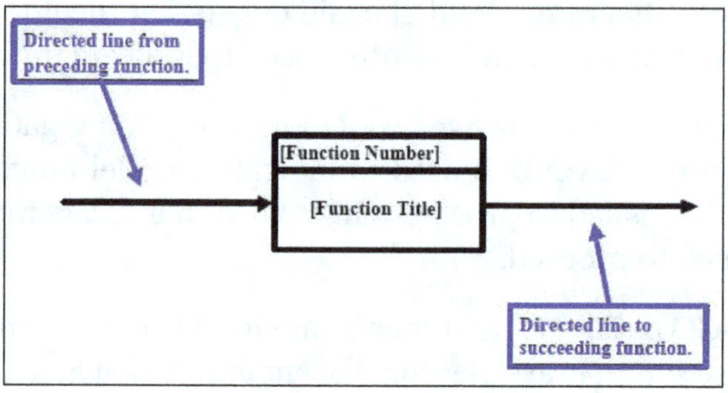

Directed Lines.

Logic Symbols

The following basic logic symbols shall be used:

- AND: A condition in which all preceding or succeeding paths are required. The symbol may contain a single input with multiple outputs or multiple inputs with a single output, but not multiple inputs and outputs combined. Read the figure as follows: F2 AND F3 may begin in parallel after completion of F1. Likewise, F4 may begin after completion of F2 AND F3.

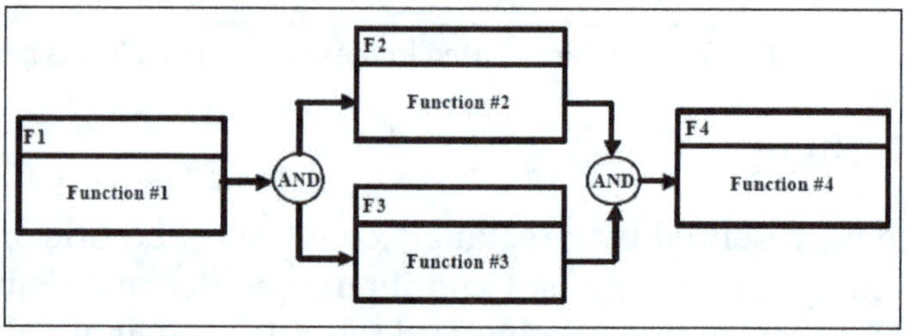

"AND" Symbol.

- Exclusive OR: A condition in which one of multiple preceding or succeeding paths is required, but not all. The symbol may contain a single input with multiple outputs or multiple inputs with single output, but not multiple inputs and outputs combined. Read the figure as follows: F2 OR F3 may begin after completion of F1. Likewise, F4 may begin after completion of either F2 OR F3.

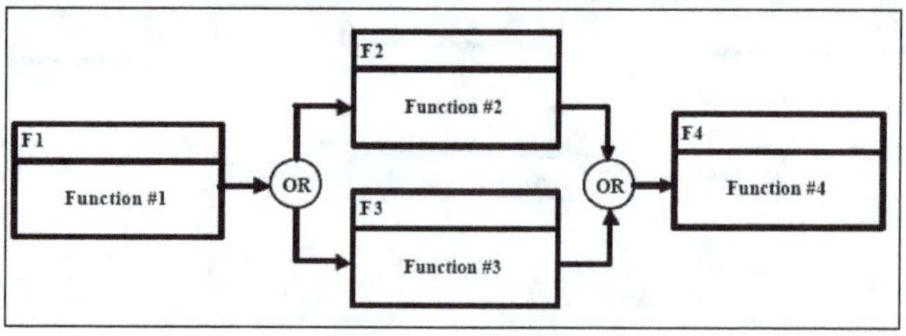

"Exclusive OR" Symbol.

- Inclusive OR: A condition in which one, some, or all of the multiple preceding or succeeding paths are required. Figure depicts Inclusive OR logic using a combination of the AND symbol and the Exclusive OR symbol. Read Figure as follows: F2 OR F3 (exclusively) may begin after completion of F1, OR (again exclusive) F2 AND F3 may begin after completion of F1. Likewise, F4 may begin after completion of either F2 OR F3 (exclusively), OR (again exclusive) F4 may begin after completion of both F2 AND F3.

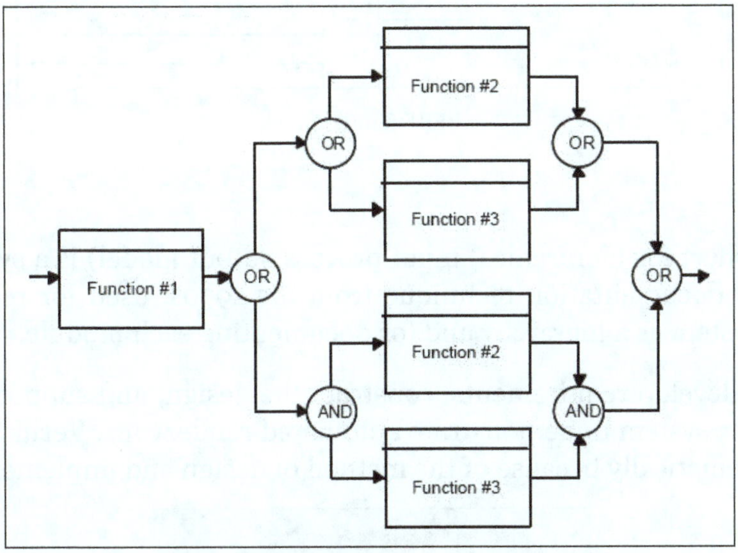

"Inclusive OR" Logic.

Contextual and Administrative Data

Each FFBD shall contain the following contextual and administrative data:

- Date the diagram was created.
- Name of the engineer, organization, or working group that created the diagram.
- Unique decimal delimited number of the function being diagrammed.
- Unique function name of the function being diagrammed.

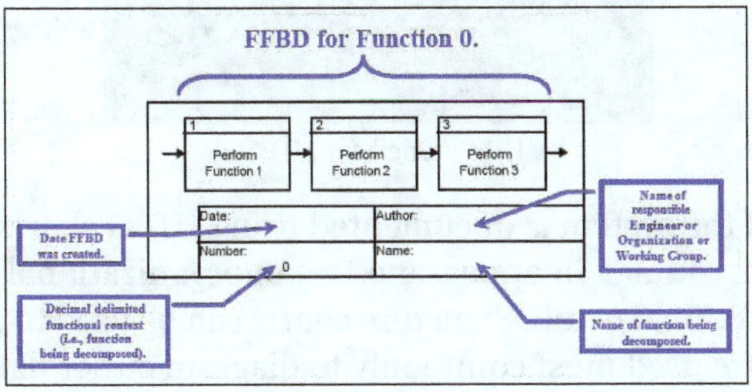

FFBD Function 0.

Figure above and below present the data in an FFBD. Figure is a decomposition of the function F2 contained in Figure and illustrates the context between functions at different levels of the model.

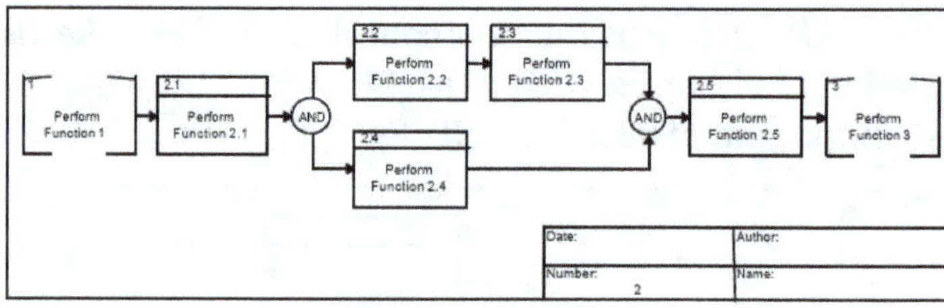

FFBD Function 2

HIPO Model

HIPO model (short for hierarchical input process output model) is a systems analysis design aid and documentation technique from the 1970s, used for representing the modules of a system as a hierarchy and for documenting each module.

It was used to develop requirements, construct the design, and support implementation of an expert system to demonstrate automated rendezvous. Verification was then conducted systematically because of the method of design and implementation.

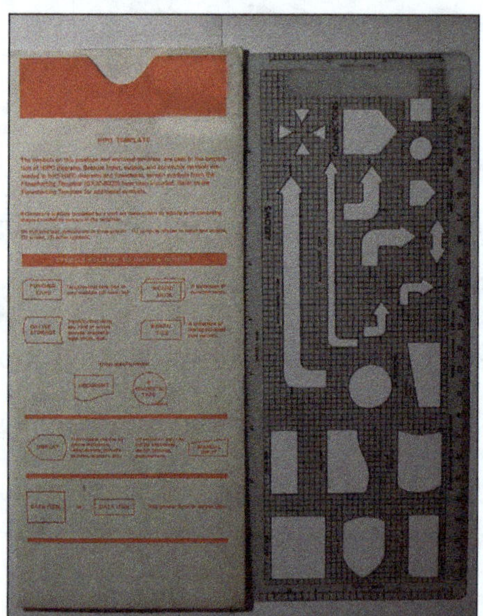

IBM stencil for HIPO.

The overall design of the system is documented using HIPO charts or structure charts. The structure chart is similar in appearance to an organizational chart, but has been modified to show additional detail. Structure charts can be used to display several types of information, but are used most commonly to diagram either data structures or code structures.

N² Diagram

The N² (pronounced "N-squared") diagram represents the logical data flow for a system or system segment. Part of the behavioral (logical architecture) representation set, the N² diagram displays the data dimension of the behavior model and helps focus attention on this subset of the model. In particular, this is helpful in partitioning and allocating the system behavior to manage internal and external interfaces.

The N² diagram is available for elements in the Function class (as well as any other subclasses of ProcessingUnit).

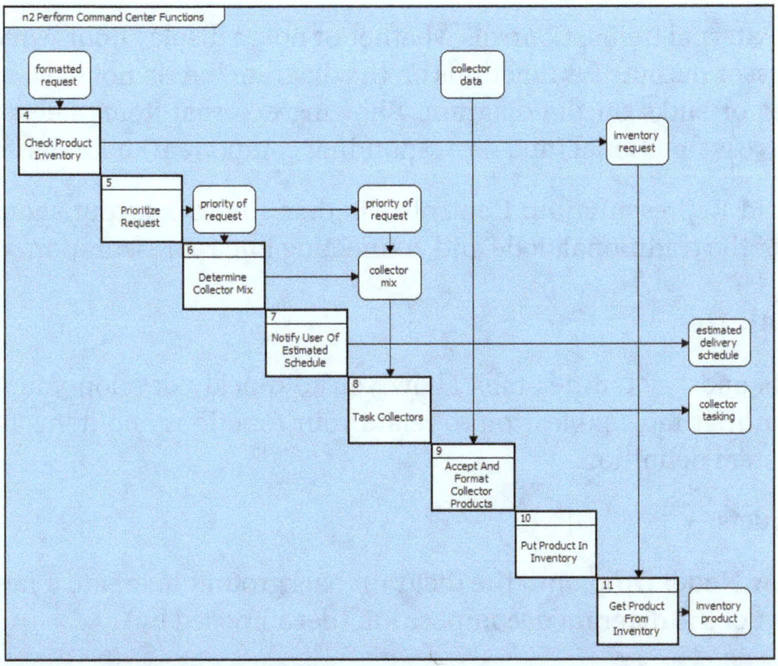

On a functional N² diagram, the subfunctions are shown on the main diagonal forming an N x N matrix of cells. Items that are output from a function are shown in the function's row. Items that are input to or trigger a function are shown in the function's column. If multiple items are output from and input to/trigger the same pair of functions, multiple items will be shown in the same item cell. If no items are exchanged between a pair of functions, the item cell will be empty.

CORE's N² diagram has been extended to display external inputs and outputs which represent external interfaces for this function. Items appearing in the top row are inputs/triggers for the function that either (i) are output by a function not displayed on this diagram or (ii) are not output by any function at all. Similarly, items in the right-hand column are outputs that are either (i) input to/trigger a function not displayed on this diagram or (ii) are not input to/trigger any function in the system model. This extension of classic N² diagrams provides valuable context but can be toggled off, as desired.

A functional N^2 diagram displays only the inputs, triggers, and outputs at that level (inputs, triggers, and outputs lower in the decomposition are not displayed). However, the decomposition of functions is considered when placing an item in a cell. For example, if an item is output from Function ABC and input to a descendent of Function XYZ, the item will be shown connecting Function ABC and Function XYZ rather than being shown as an external output.

Diagram Options

In addition to the classic diagram options, the N2 diagram settings include:

- Show External Items: Controls whether or not to display items which are inputs, triggers, or outputs for functions on the diagram but do not have corresponding sources or sinks on the diagram. Showing external items helps identify data flow across functional (and corresponding component) boundaries.
- Use Grid Representation: Controls whether a matrix layout should be used in place of the traditional node and connecting line representation.

Diagram Palette

The constructs and key entities tabs allow you to quickly develop your N2 diagram, while the all entities tab enables you to relate your functions and items to the remainder of your system definition.

- Constructs:
    - New Node: Drop onto the diagram background to create a new function as part of the diagram decomposition (decomposed by).
    - Nodes: Drop onto the diagram background to add an existing function as part of the diagram decomposition (decomposed by).
    - New Input: Drop onto any function node to create a new item that is input to the node.
    - Inputs: Drop onto any function node to relate an existing item that is input to the node.
    - New Trigger: Drop onto any function node to create a new item that triggers the node.
    - Triggers: Drop onto any function node to relate an existing item that triggers the node.
    - New Output: Drop onto any function node to create a new item that is output from the node.

- Outputs: Drop onto any function node to relate an existing item that is output from the node.
- New Information Block: Drop onto the diagram to insert an information block (a mini property sheet).
- New Note: Drop onto the diagram to insert a new note (descriptive text in a note icon).
- New Shape: Drop onto the diagram to insert a new shape (a rectangle, rounded rectangle, circle, or ellipse with text, if desired).
- New Graphic: Drop onto the diagram to insert a new picture.

- Key Entities:
  - Function: Drop an existing function onto the diagram background to it as part of the diagram decomposition (decomposed by).
  - Item: Drop an existing item onto a function node to relate it to the node using the input to, triggers, or output from relation.
- All Entities: All classes and elements in the system model, allowing you to drag any element on top of a diagram node to establish relationships with the balance of your system model.

References

- It-systems-development: fca.gov, Retrieved 02 February, 2020
- Systems-development-life-cycle: searchsoftwarequality.techtarget.com, Retrieved 18 June, 2020
- Land, S.K.; Smith, D.B.; Walz, J.W. (2012). Practical Support for Lean Six Sigma Software Process Definition: Using IEEE Software Engineering Standards. John Wiley & Sons. pp. 341–3. ISBN 9780470289952
- Suh (2001). Axiomatic Design: Advances and Applications, Oxford University Press, 2001, ISBN 0-19-513466-4
- Katzan, Harry, Jr. Systems Design and Documentation: An introduction to the HIPO Method. Van Nostrand Reinhold. ISBN 0-442-24267-0

4
Systems Modeling Languages

Systems modeling language is referred to as a general purpose modeling language that assists the analysis, design and validation of a wide range of systems. Various diagrams are studied under it such as package diagram, use case diagram, activity diagram, requirements diagram, state diagram, etc. This chapter closely examines the subject of systems modeling languages to provide an easy understanding of the subject.

The Systems Modelling Language (SysML) is a general purpose modelling language for engineering systems. SysML supports the analysis, design and verification of complex systems including hardware, software, information, personnel, procedures, and facilities in a graphical notation. SysML provides graphical representations based on a semantic foundation for modelling system behaviour, requirements, structure, and parametrical elements which are used to integrate with other engineering analysis models using the OML XML Metadata Interchange (XMI®).

SysML is defined as an extension of a subset of the Unified Modelling Language (UML) using UML's profile mechanism and was submitted to the Object Management Group (OMG) in November 2005. A number of competing specification proposals were merged and adopted by the OMG as OMG SysML.

Relationship between SysML and UML

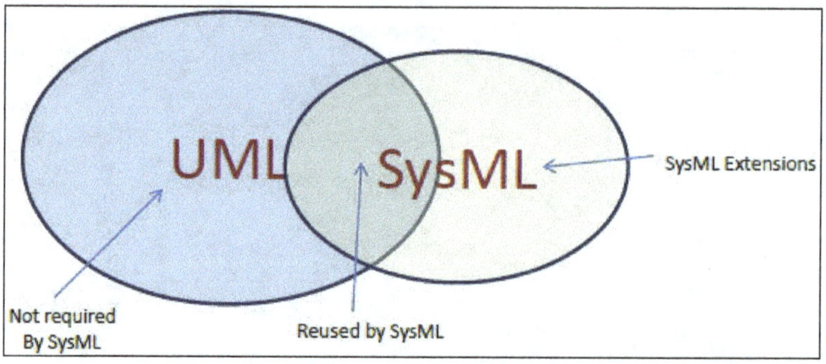

SysML uses the concept of the 'block' as the basic unit structure to represent hardware, software, information, personnel, procedures, and facilities. Block definition diagrams and internal block diagrams are used to represent a system structure.

SysML Diagram Types

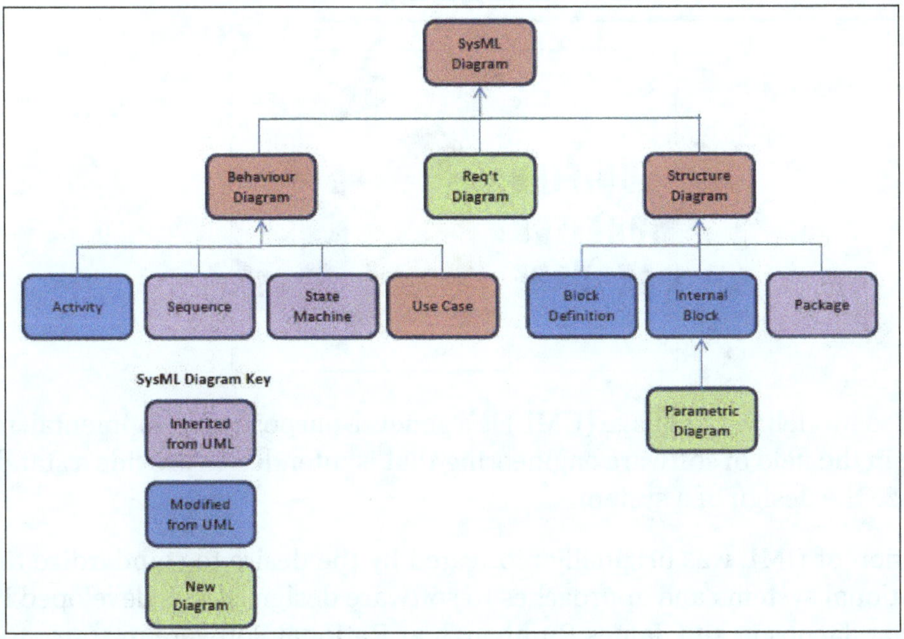

The SysML Behaviour Diagram includes:

- SysML Use Case Diagram: High level description of functionality delivered through interactions between parts of the system.
- SysML Activity Diagram: Illustrates the flow of data and control between activities.
- SysML Sequence Diagram: Interactions between collaboration parts of the system.
- SysML State Machine Diagram: Identifies transitions occurring in response to events in the system.

The SysML Requirements Diagram identifies requirements hierarchies and derivation. Satisfy and verify relationships allow a requirement to be related to a model element that satisfies or verifies the requirements. This diagram acts as the connection between typical requirements management tools and system models.

The SysML Parametric Diagram identifies constraints on system property values such as performance, reliability and mass properties. It provides the ability to integrate the specification and design models with engineering analysis models.

An Allocation Relationship is also supported within SysML which represents allocating functions to components, logical to physical components, and software to hardware.

Unified Modeling Language

The Unified Modeling Language (UML) is a general-purpose, developmental, modeling language in the field of software engineering that is intended to provide a standard way to visualize the design of a system.

The creation of UML was originally motivated by the desire to standardize the disparate notational systems and approaches to software design. It was developed by Grady Booch, Ivar Jacobson and James Rumbaugh at Rational Software in 1994–1995, with further development led by them through 1996.

In 1997, UML was adopted as a standard by the Object Management Group (OMG), and has been managed by this organization ever since. In 2005, UML was also published by the International Organization for Standardization (ISO) as an approved ISO standard. Since then the standard has been periodically revised to cover the latest revision of UML.

Before UML 1.0

UML has been evolving since the second half of the 1990s and has its roots in the object-oriented programming methods developed in the late 1980s and early 1990s. The timeline shows the highlights of the history of object-oriented modeling methods and notation.

It is originally based on the notations of the Booch method, the object-modeling technique (OMT) and object-oriented software engineering (OOSE), which it has integrated into a single language.

Rational Software Corporation hired James Rumbaugh from General Electric in 1994 and after that the company became the source for two of the most popular object-oriented modeling approaches of the day: Rumbaugh's object-modeling technique (OMT) and Grady Booch's method. They were soon assisted in their efforts by Ivar Jacobson, the creator of the object-oriented software engineering (OOSE) method, who joined them at Rational in 1995.

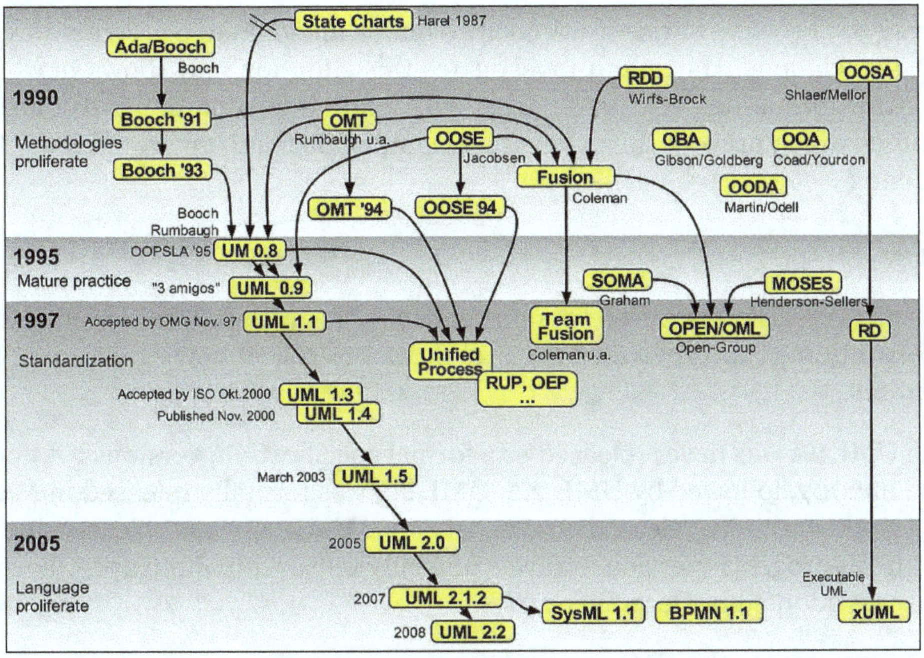

Object-oriented methods and notation.

UML 1.x

Under the technical leadership of those three (Rumbaugh, Jacobson and Booch), a consortium called the UML Partners was organized in 1996 to complete the Unified Modeling Language (UML) specification, and propose it to the Object Management Group (OMG) for standardisation. The partnership also contained additional interested parties (for example HP, DEC, IBM and Microsoft). The UML Partners' UML 1.0 draft was proposed to the OMG in January 1997 by the consortium. During the same month the UML Partners formed a group, designed to define the exact meaning of language constructs, chaired by Cris Kobryn and administered by Ed Eykholt, to finalize the specification and integrate it with other standardization efforts. The result of this work, UML 1.1, was submitted to the OMG in August 1997 and adopted by the OMG in November 1997.

After the first release a task force was formed to improve the language, which released several minor revisions, 1.3, 1.4, and 1.5. The standards it produced (as well as the original standard) have been noted as being ambiguous and inconsistent.

Cardinality Notation

As with database Chen, Bachman, and ISO ER diagrams, class models are specified to use "look-across" cardinalities, even though several authors prefer same-side or "look-here" for roles and both minimum and maximum cardinalities. The "look-across" technique used by UML and ER diagrams is less effective and less coherent when applied to *n*-ary relationships of order strictly greater than 2.

Feinerer says: "Problems arise if we operate under the look-across semantics as used for UML associations. Hartmann investigates this situation and shows how and why different transformations fail.", and: "The look-across interpretation introduces several difficulties which prevent the extension of simple mechanisms from binary to *n*-ary associations."

UML 2

UML 2.0 major revision replaced version 1.5 in 2005, which was developed with an enlarged consortium to improve the language further to reflect new experience on usage of its features.

Although UML 2.1 was never released as a formal specification, versions 2.1.1 and 2.1.2 appeared in 2007, followed by UML 2.2. UML 2.3 was formally released in May 2010. UML 2.4.1 was formally released in August 2011. UML 2.5 was released in October 2012 as an "In progress" version and was officially released in June 2015. Formal version 2.5.1 was adopted in December 2017.

There are four parts to the UML 2.x specification:

- The Superstructure that defines the notation and semantics for diagrams and their model elements.
- The Infrastructure that defines the core metamodel on which the Superstructure is based.
- The Object Constraint Language (OCL) for defining rules for model elements.
- The UML Diagram Interchange that defines how UML 2 diagram layouts are exchanged.

The current versions of these standards are:

- UML Superstructure version 2.4.1.
- UML Infrastructure version 2.4.1.
- OCL version 2.3.1.
- UML Diagram Interchange version 1.0.

It continues to be updated and improved by the revision task force, who resolve any issues with the language.

Design

UML offers a way to visualize a system's architectural blueprints in a diagram, including elements such as:

- Any activities (jobs).
- Individual components of the system; and how they can interact with other software components.
- How the system will run.
- How entities interact with others (components and interfaces).
- External user interface.

Although originally intended for object-oriented design documentation, UML has been extended to a larger set of design documentation, and been found useful in many contexts.

Software Development Methods

UML is not a development method by itself; however, it was designed to be compatible with the leading object-oriented software development methods of its time, for example OMT, Booch method, Objectory and especially RUP that it was originally intended to be used with when work began at Rational Software.

Modeling

It is important to distinguish between the UML model and the set of diagrams of a system. A diagram is a partial graphic representation of a system's model. The set of diagrams need not completely cover the model and deleting a diagram does not change the model. The model may also contain documentation that drives the model elements and diagrams (such as written use cases).

UML diagrams represent two different views of a system model:

- Static (or structural) view: Emphasizes the static structure of the system using objects, attributes, operations and relationships. It includes class diagrams and composite structure diagrams.
- Dynamic (or behavioral) view: Emphasizes the dynamic behavior of the system by showing collaborations among objects and changes to the internal states of objects. This view includes sequence diagrams, activity diagrams and state machine diagrams.

UML models can be exchanged among UML tools by using the XML Metadata Interchange (XMI) format. In UML, one of the key tools for behavior modelling is the use-case model, caused by OOSE. Use cases are a way of specifying required usages of a system. Typically, they are used to capture the requirements of a system, that is, what a system is supposed to do.

Diagrams

UML 2 has many types of diagrams, which are divided into two categories. Some types represent structural information, and the rest represent general types of behavior, including a few that represent different aspects of interactions. These diagrams can be categorized hierarchically as shown in the following class diagram:

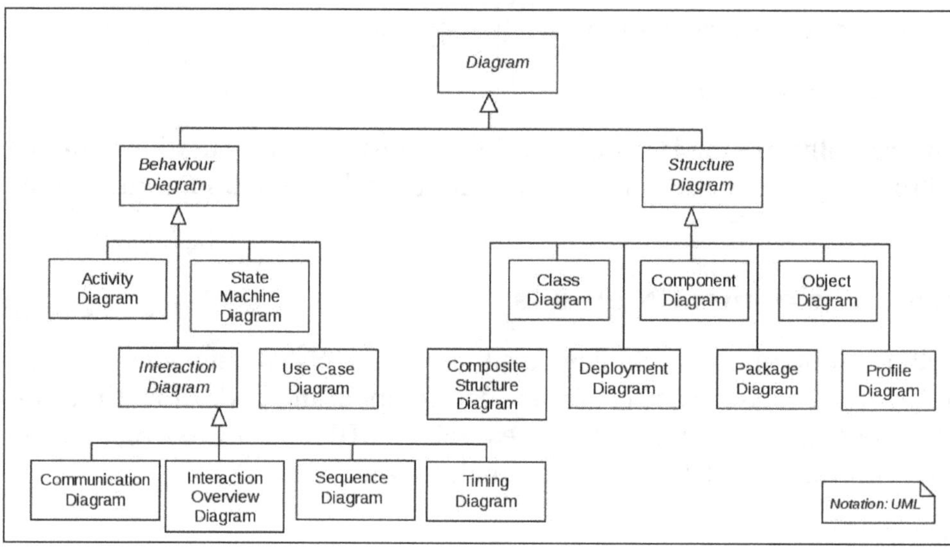

Structure Diagrams

Structure diagrams emphasize the things that must be present in the system being modeled. Since structure diagrams represent the structure, they are used extensively in documenting the software architecture of software systems. For example, the component diagram describes how a software system is split up into components and shows the dependencies among these components.

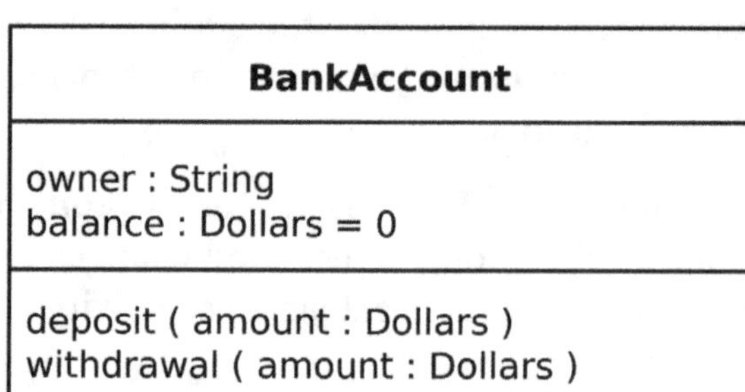

Class diagram.

Behavior Diagrams

Behavior diagrams emphasize what must happen in the system being modeled. Since behavior diagrams illustrate the behavior of a system, they are used extensively to describe the functionality of software systems. As an example, the activity diagram describes the business and operational step-by-step activities of the components in a system.

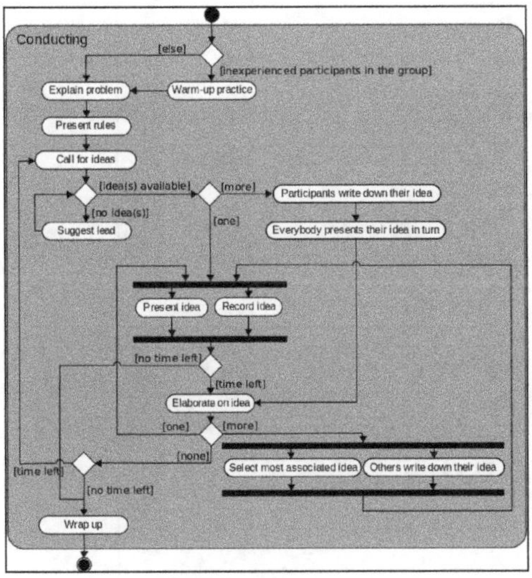

Activity diagram.

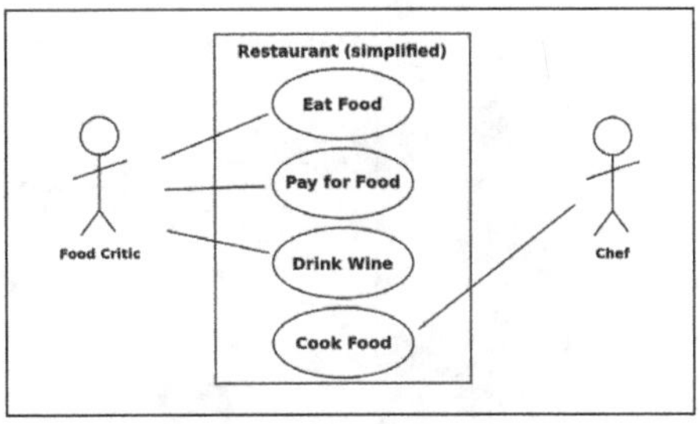

Use case diagram.

Interaction Diagrams

Interaction diagrams, a subset of behavior diagrams, emphasize the flow of control and data among the things in the system being modeled. For example, the sequence diagram shows how objects communicate with each other regarding a sequence of messages.

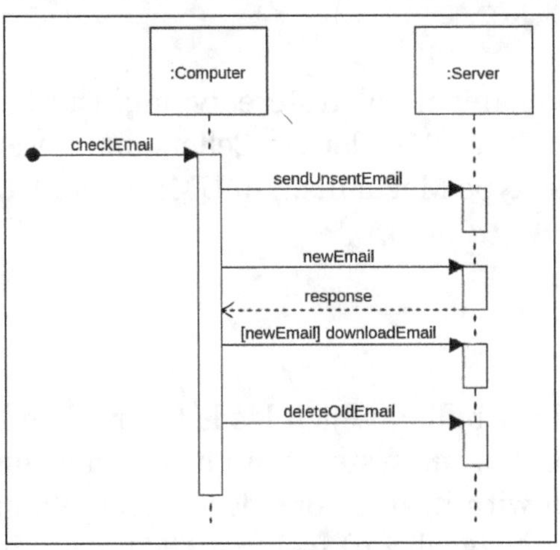

Sequence diagram.

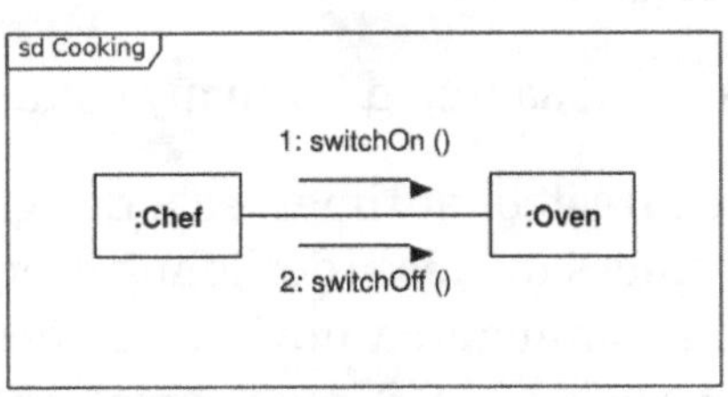

Communication diagram.

Metamodeling

The Object Management Group (OMG) has developed a metamodeling architecture to define the UML, called the Meta-Object Facility. MOF is designed as a four-layered architecture, as shown in the image at right. It provides a meta-meta model at the top, called the M3 layer. This M3-model is the language used by Meta-Object Facility to build metamodels, called M2-models.

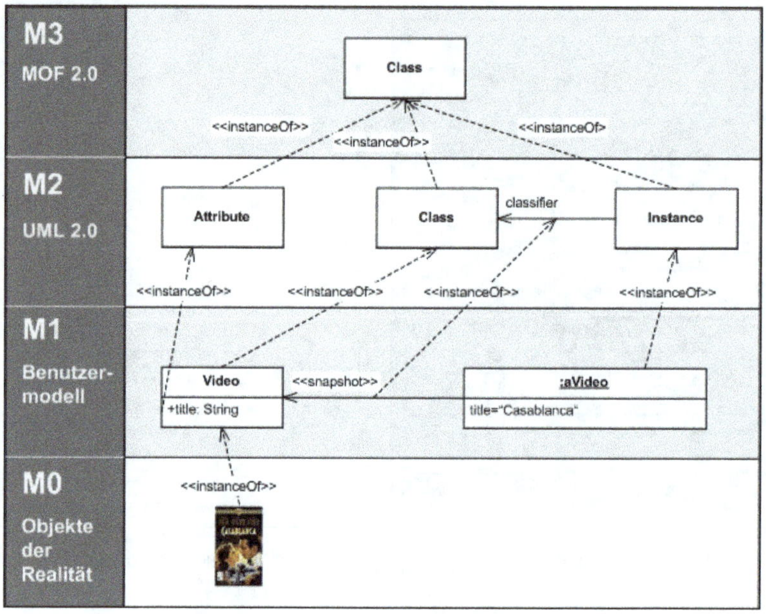

Illustration of the Meta-Object Facility.

The most prominent example of a Layer 2 Meta-Object Facility model is the UML metamodel, which describes the UML itself. These M2-models describe elements of the M1-layer, and thus M1-models. These would be, for example, models written in UML. The last layer is the M0-layer or data layer. It is used to describe runtime instances of the system.

The meta-model can be extended using a mechanism called stereotyping. This has been criticised as being insufficient/untenable by Brian Henderson-Sellers and Cesar Gonzalez-Perez in "Uses and Abuses of the Stereotype Mechanism in UML 1.x and 2.0".

Adoption

UML has been marketed for many contexts.

It has been treated, at times, as a design silver bullet, which leads to problems. UML misuse includes overuse (designing every part of the system with it, which is unnecessary) and assuming that novices can design with it. It is considered a large language, with many constructs. Some people (including Jacobson) feel that UML's size hinders learning (and therefore, using) it.

System Context Diagram

A system context diagram (SCD) in engineering is a diagram that defines the boundary between the system, or part of a system, and its environment, showing the entities that interact with it. This diagram is a high level view of a system. It is similar to a block diagram.

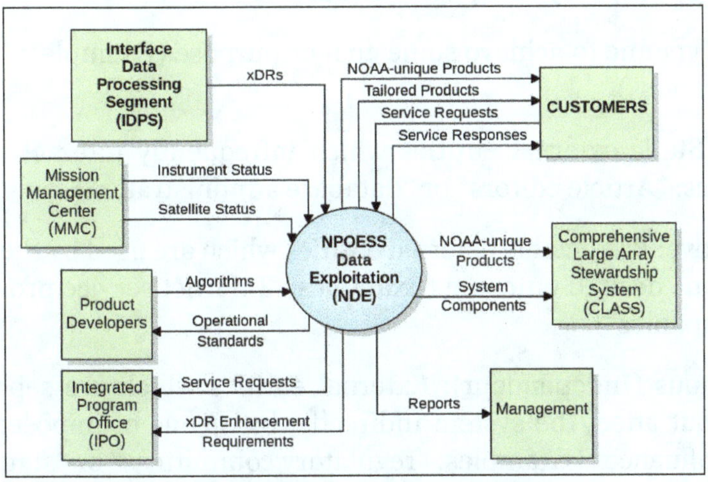

Example of a system context diagram.

System context diagrams show a system, as a whole and its inputs and outputs from/to external factors. According to Kossiakoff and Sweet:

> "System Context Diagrams represent all external entities that may interact with a system Such a diagram pictures the system at the center, with no details of its interior structure, surrounded by all its interacting systems, environments and activities. The objective of the system context diagram is to focus attention on external factors and events that should be considered in developing a complete set of systems requirements and constraints."

System context diagrams are used early in a project to get agreement on the scope under investigation. Context diagrams are typically included in a requirements document. These diagrams must be read by all project stakeholders and thus should be written in plain language, so the stakeholders can understand items within the document.

Building Blocks

Context diagrams can be developed with the use of two types of building blocks:

- Entities (Actors): Labeled boxes; one in the center representing the system, and around it multiple boxes for each external actor.
- Relationships: Labeled lines between the entities and system.

For example, "customer places order." Context diagrams can also use many different drawing types to represent external entities. They can use ovals, stick figures, pictures, clip art or any other representation to convey meaning. Decision trees and data storage are represented in system flow diagrams.

A context diagram can also list the classifications of the external entities as one of a set of simple categories which add clarity to the level of involvement of the entity with regards to the system. These categories include:

- Active: Dynamic to achieve some goal or purpose (Examples: "Article readers" or "customers").
- Passive: Static external entities which infrequently interact with the system (Examples: "Article editors" or "database administrator").
- Cooperative: Predictable external entities which are used by the system to bring about some desired outcome (Examples: "Internet service providers" or "shipping companies").
- Autonomous (Independent): External entities which are separated from the system, but affect the system indirectly, by means of imposed constraints or similar influences (Examples: "regulatory committees" or "standards groups").

Alternatives

The best system context diagrams are used to display how a system interoperates at a very high level, or how systems operate and interact logically. The system context diagram is a necessary tool in developing a baseline interaction between systems and actors; actors and a system or systems and systems. Alternatives to the system context diagram are:

- Architecture Interconnect Diagram: The figure gives an example of an Architecture Interconnect Diagram. A representation of the Albuquerque regional ITS architecture interconnects for the Albuquerque Police Department that was generated using the Turbo Architecture tool is shown in the figure. Each block represents an ITS inventory element, including the name of the stakeholder in the top shaded portion. The interconnect lines between elements are solid or dashed, indicating existing or planned connections.
- Business Model Canvas, a strategic management template for developing new or documenting existing business models. It is a visual chart with elements describing a firm's value proposition, infrastructure, customers, and finances. It assists firms in aligning their activities by illustrating potential trade-offs.
- Enterprise data model: This type of data model according to Simsion can contain up to 50 to 200 entity classes, which results from specific "high level of

generalization in data modeling".

- IDEF0 Top Level Context Diagram: The IDEF0 process starts with the identification of the prime function to be decomposed. This function is identified on a "Top Level Context Diagram" that defines the scope of the particular IDEF0 analysis.
- Problem Diagrams (Problem Frames): In addition to the kinds of things shown on a context diagram, a problem diagram shows requirements and requirements references.
- Use case diagram: One of the Unified Modeling Language diagrams. They also represent the scope of the project at a similar level of abstraction. - Use Cases, however, tend to focus more on the goals of 'actors' who interact with the system, and do not specify any solution. Use Case diagrams represent a set of Use Cases, which are textual descriptions of how an actor achieves the goal of a use case.
- ArchiMate: ArchiMate is an open and independent enterprise architecture modeling language to support the description, analysis and visualization of architecture within and across business domains in an unambiguous way.

Most of these diagrams work well as long as a limited number of interconnects will be shown. Where twenty or more interconnects must be displayed, the diagrams become quite complex and can be difficult to read.

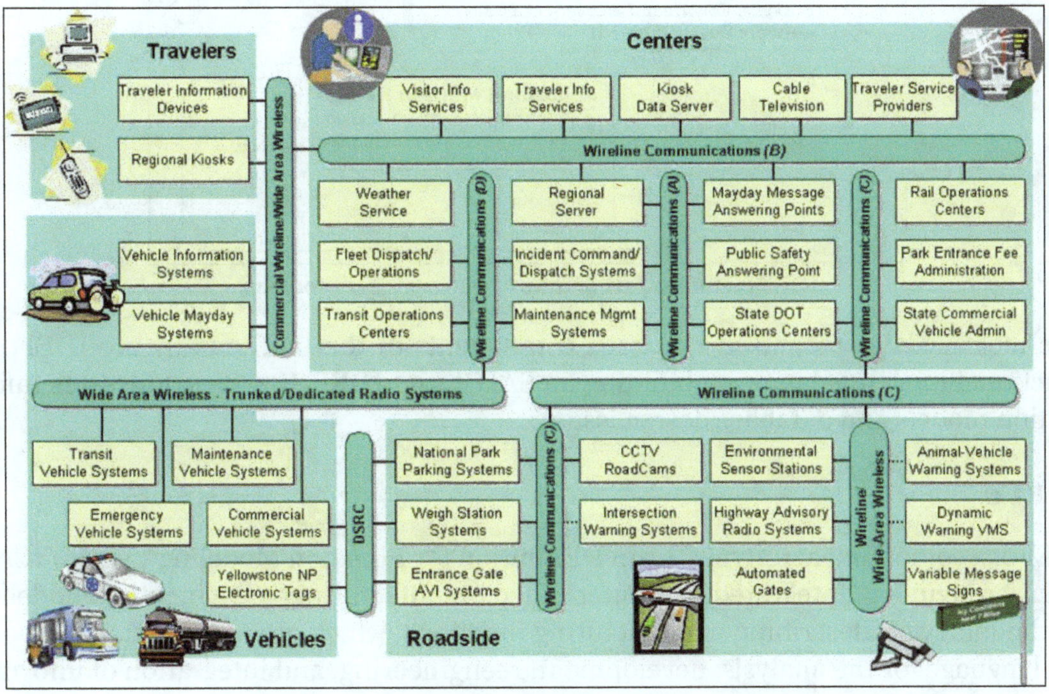

Example of an Architecture Interconnect Diagram.

IDEF

IDEF, initially an abbreviation of ICAM Definition and renamed in 1999 as Integration Definition, is a family of modeling languages in the field of systems and software engineering. They cover a wide range of uses from functional modeling to data, simulation, object-oriented analysis and design, and knowledge acquisition. These definition languages were developed under funding from U.S. Air Force and, although still most commonly used by them and other military and United States Department of Defense (DoD) agencies, are in the public domain.

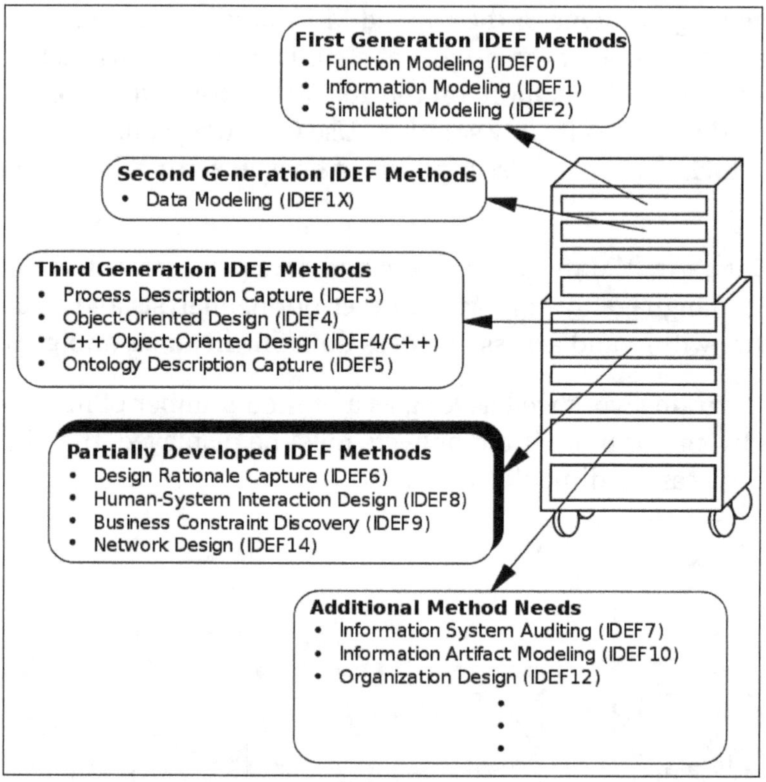

IDEF methods: part of the systems engineer's toolbox.

The most-widely recognized and used components of the IDEF family are IDEF0, a functional modeling language building on SADT, and IDEF1X, which addresses information models and database design issues.

IDEF0

IDEF0, a compound acronym ("Icam DEFinition for Function Modeling", where ICAM is an acronym for "Integrated Computer Aided Manufacturing"), is a function modeling methodology for describing manufacturing functions, which offers a functional modeling language for the analysis, development, reengineering, and integration of information systems; business processes; or software engineering analysis.

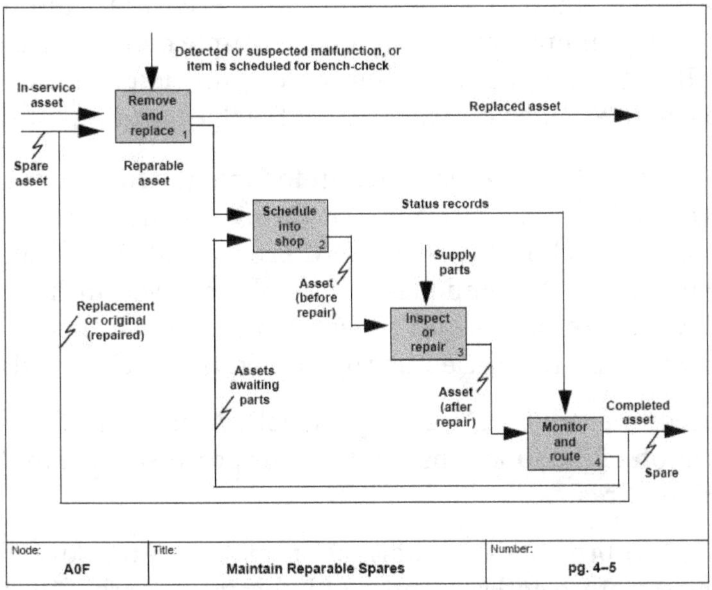

IDEFo Diagram Example.

IDEFo is part of the IDEF family of modeling languages in the field of software engineering, and is built on the functional modeling language Structured Analysis and Design Technique (SADT).

The IDEFo Functional Modeling method is designed to model the decisions, actions, and activities of an organization or system. It was derived from the established graphic modeling language Structured Analysis and Design Technique (SADT) developed by Douglas T. Ross and SofTech, Inc. In its original form, IDEFo includes both a definition of a graphical modeling language (syntax and semantics) and a description of a comprehensive methodology for developing models. The US Air Force commissioned the SADT developers "to develop a function model method for analyzing and communicating the functional perspective of a system. IDEFo should assist in organizing system analysis and promote effective communication between the analyst and the customer through simplified graphical devices".

Where the Functional flow block diagram is used to show the functional flow of a product, IDEFo is used to show data flow, system control, and the functional flow of lifecycle processes. IDEFo is capable of graphically representing a wide variety of business, manufacturing and other types of enterprise operations to any level of detail. It provides rigorous and precise description, and promotes consistency of usage and interpretation. It is well-tested and proven through many years of use by government and private industry. It can be generated by a variety of computer graphics tools. Numerous commercial products specifically support development and analysis of IDEFo diagrams and models.

An associated technique, Integration Definition for Information Modeling (IDEF1x), is used to supplement IDEFo for data-intensive systems. The IDEFo standard, Federal

Information Processing Standards Publication 183 (FIPS 183), and the IDEF1x standard (FIPS 184) are maintained by the National Institute of Standards and Technology (NIST). FIPS PUB 183 "Integration Definition for Function Modeling (IDEF0)," was withdrawn as a Federal Standard (in favor of OPEN Specifications and Standards).

During the 1970s, the U.S. Air Force Program for Integrated Computer Aided Manufacturing (ICAM) sought to increase manufacturing productivity through systematic application of computer technology. The ICAM program identified the need for better analysis and communication techniques for people involved in improving manufacturing productivity. As a result, in 1981 the ICAM program developed a series of techniques known as the IDEF (ICAM Definition) techniques which included the following:

- IDEF0, used to produce a "function model": A function model is a structured representation of the functions, activities or processes within the modeled system or subject area.
- IDEF1, used to produce an "information model": An information model represents the structure and semantics of information within the modeled system or subject area.
- IDEF2, used to produce a "dynamics model": A dynamics model represents the time-varying behavioral characteristics of the modeled system or subject area.

In 1983, the U.S. Air Force Integrated Information Support System program enhanced the IDEF1 information modeling technique to form IDEF1X (IDEF1 Extended), a semantic data modeling technique. By the 1990s, IDEF0 and IDEF1X techniques are widely used in the government, industrial and commercial sectors, supporting modeling efforts for a wide range of enterprises and application domains. In 1991 the National Institute of Standards and Technology (NIST) received support from the U.S. Department of Defense, Office of Corporate Information Management (DoD/CIM), to develop one or more Federal Information Processing Standard (FIPS) for modeling techniques. The techniques selected were IDEF0 for function modeling and IDEF1X for information modeling. These FIPS documents are based on the IDEF manuals published by the U.S. Air Force in the early 1980s.

IDEF0 Topics

The IDEF0 Approach

IDEF0 may be used to model a wide variety of automated and non-automated systems. For new systems, it may be used first to define the requirements and specify the functions, and then to design an implementation that meets the requirements and performs the functions. For existing systems, IDEF0 can be used to analyze the functions the system performs and to record the mechanisms (means) by which these are done. The result of applying IDEF0 to a system is a model that consists of a hierarchical series of diagrams, text, and glossary cross-referenced to each other. The two primary modeling components are functions (represented on a diagram by boxes) and the data and objects that inter-relate those functions (represented by arrows).

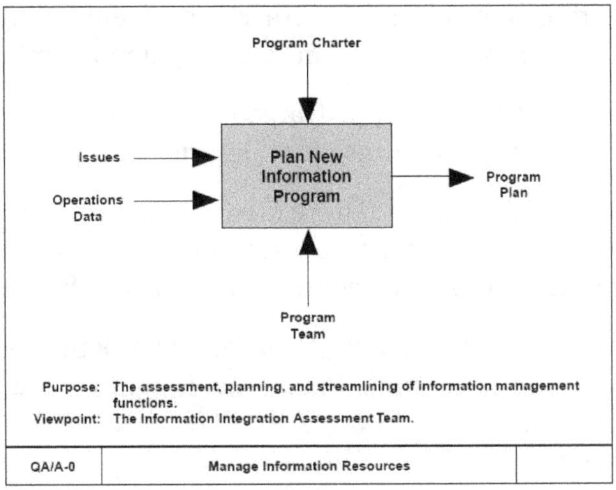

Top-Level Context Diagram.

IDEFo Building Blocks

The IDEFo model displayed here on the left is based on a simple syntax. Each activity is described by a verb-based label placed in a box. Inputs are shown as arrows entering the left side of the activity box while output are shown as exiting arrows on the right side of the box. Controls are displayed as arrows entering the top of the box and mechanisms are displayed as arrows entering from the bottom of the box. Inputs, Controls, Outputs, and Mechanisms (ICOM) are all referred to as concepts.

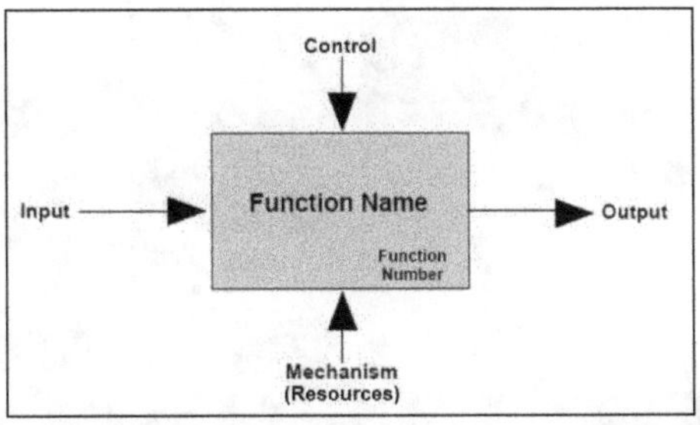

Integration Definition for Function Modeling (IDEFo) Box Format.

- Arrow: A directed line, composed of one or more arrow segments, that models an open channel or conduit conveying data or objects from source (no arrowhead) to use (with arrowhead). There are 4 arrow classes: Input Arrow, Output Arrow, Control Arrow, and Mechanism Arrow (includes Call Arrow).
- Box: A rectangle, containing a name and number, used to represent a function.
- Context: The immediate environment in which a function (or set of functions on a diagram) operates.

- Decomposition: The partitioning of a modeled function into its component functions.
- Fork: The junction at which an IDEFo arrow segment (going from source to use) divides into two or more arrow segments. May denote unbundling of meaning.
- Function: An activity, process, or transformation (modeled by an IDEFo box) identified by a verb or verb phrase that describes what must be accomplished.
- Join: The junction at which an IDEFo arrow segment (going from source to use) merges with one or more other arrow segments to form a single arrow segment. May denote bundling of arrow segment meanings.
- Node: A box from which child boxes originate; a parent box. See Node Index, Node Tree, Node Number, Node Reference, Diagram Node Number.

Graphical Notation

IDEFo is a model that consists of a hierarchical series of diagrams, text, and glossary cross referenced to each other. The two primary modeling components are:

- Functions (represented on a diagram by boxes).
- Data and objects that interrelate those functions (represented by arrows).

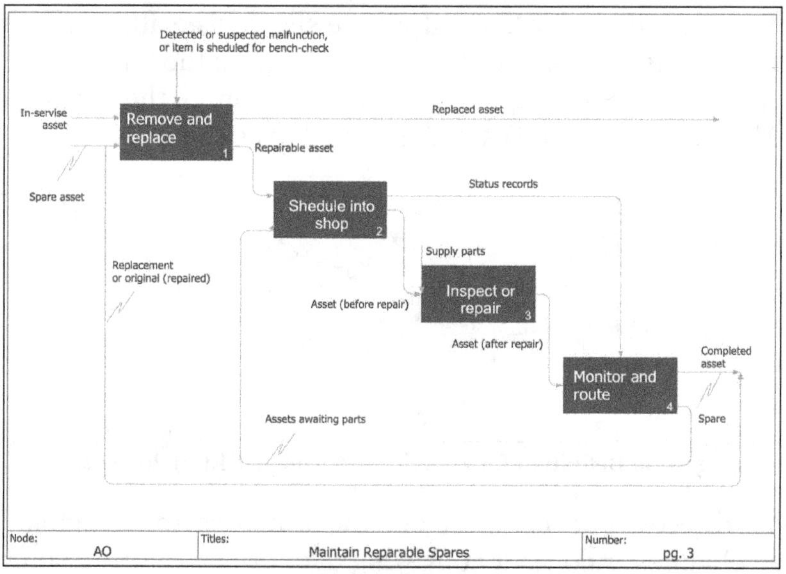

IDEFo Diagram Example.

As shown by Figure the position at which the arrow attaches to a box conveys the specific role of the interface. The controls enter the top of the box. The inputs, the data or objects acted upon by the operation, enter the box from the left. The outputs of the operation leave the right-hand side of the box. Mechanism arrows that provide supporting means for performing the function join (point up to) the bottom of the box.

The IDEF0 Process

The IDEF0 process starts with the identification of the prime function to be decomposed. This function is identified on a "Top Level Context Diagram," that defines the scope of the particular IDEF0 analysis. An example of a Top Level Context Diagram for an information system management process is shown in figure. From this diagram lower-level diagrams are generated. An example of a derived diagram, called a "child" in IDEF0 terminology, for a life cycle function is shown in figure.

IDEF1X

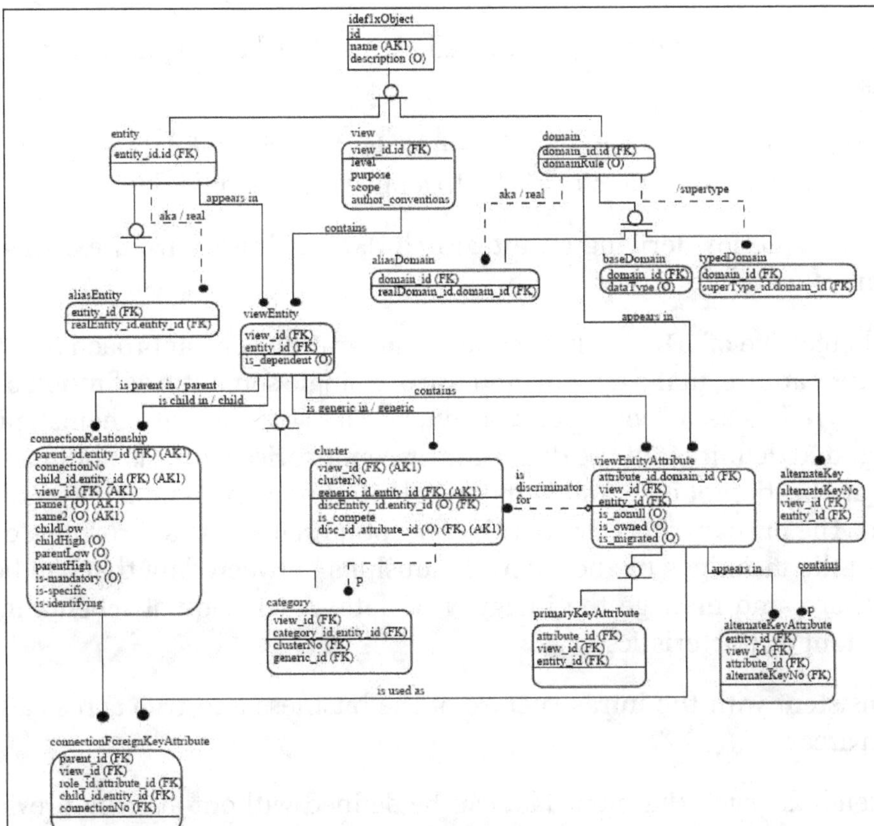

Example of an IDEF1X diagram.

Integration DEFinition for information modeling (IDEF1X) is a data modeling language for the development of semantic data models. IDEF1X is used to produce a graphical information model which represents the structure and semantics of information within an environment or system.

IDEF1X permits the construction of semantic data models which may serve to support the management of data as a resource, the integration of information systems, and the building of computer databases. This standard is part of the IDEF family of modeling languages in the field of software engineering.

A data modeling technique is used to model data in a standard, consistent and predictable manner in order to manage it as a resource. It can be used in projects requiring a standard means of defining and analyzing the data resources within an organization. Such projects include the incorporation of a data modeling technique into a methodology, managing data as a resource, integrating information systems, or designing computer databases. The primary objectives of the IDEF1X standard are to provide:

- Means for completely understanding and analyzing an organization's data resources.
- Common means of representing and communicating the complexity of data.
- A technique for presenting an overall view of the data required to run an enterprise.
- Means for defining an application-independent view of data which can be validated by users and transformed into a physical database design.
- A technique for deriving an integrated data definition from existing data resources.

A principal objective of IDEF1X is to support integration. The approach to integration focuses on the capture, management, and use of a single semantic definition of the data resource referred to as a "conceptual schema." The "conceptual schema" provides a single integrated definition of the data within an enterprise which is not biased toward any single application of data and is independent of how the data is physically stored or accessed. The primary objective of this conceptual schema is to provide a consistent definition of the meanings of and interrelationships between data that can be used to integrate, share, and manage the integrity of data. A conceptual schema must have three important characteristics:

- Consistent with the infrastructure of the business and true across all application areas.
- Extendible, such that new data can be defined without altering previously defined data.
- Transformable to both the required user views and to a variety of data storage and access structures.

IDEF1X Building Blocks

- Entities: The representation of a class of real or abstract things (people, objects, places, events, ideas, combination of things, etc.) that are recognized as instances of the same class because they share the same characteristics and can participate in the same relationships.

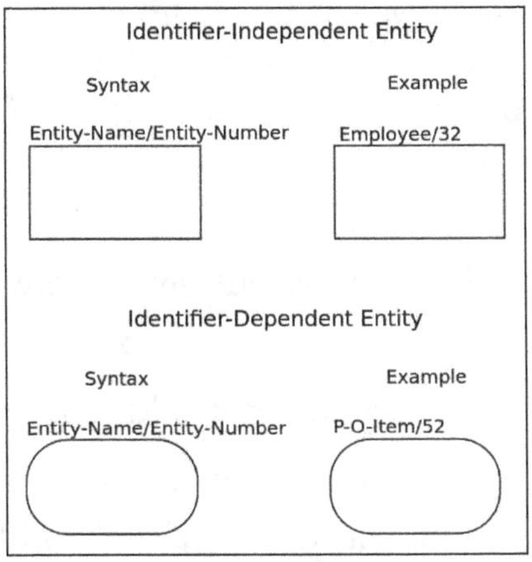

Entity syntax.

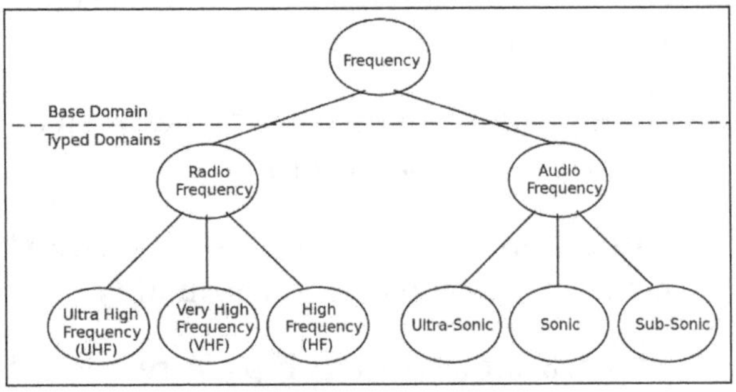

Domain hierarchy.

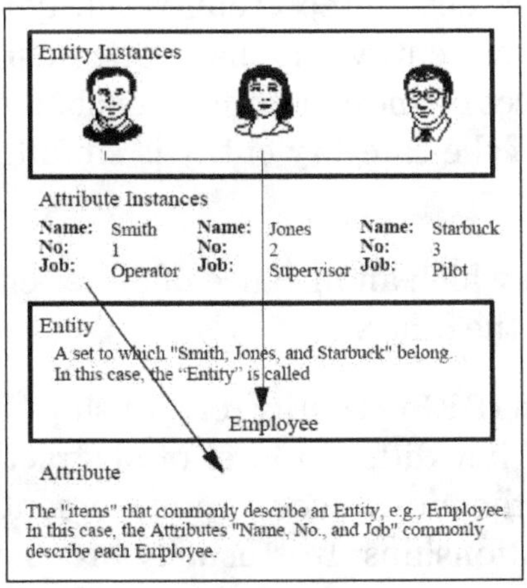

Attribute example.

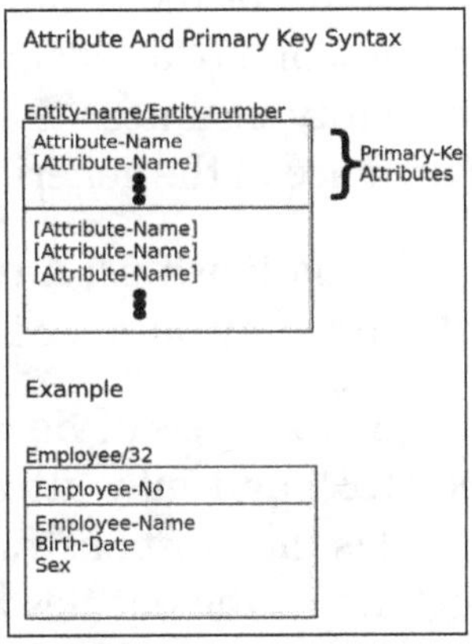

Primary key syntax.

- Domains: A named set of data values (fixed, or possibly infinite in number) all of the same data type, upon which the actual value for an attribute instance is drawn. Every attribute must be defined on exactly one underlying domain. Multiple attributes may be based on the same underlying domain.
- Attributes: A property or characteristic that is common to some or all of the instances of an entity. An attribute represents the use of a domain in the context of an entity.
- Keys: An attribute, or combination of attributes, of an entity whose values uniquely identify each entity instance. Each such set constitutes a candidate key.
- Primary keys: The candidate key selected as the unique identifier of an entity.
- Foreign keys: An attribute, or combination of attributes of a child or category entity instance whose values match those in the primary key of a related parent

or generic entity instance. A foreign key can be viewed as the result of the "migration" of the primary key of the parent or generic entity through a specific connection or categorization relationship. An attribute or combination of attributes in the foreign key can be assigned a role name reflecting its role in the child or category entity.

- Relationships: An association between the instances of two entities or between instances of the same entity.
- Connection relationships: A relationship having no semantics in addition to association. See constraint, cardinality.
- Categorization relationships: A relationship in which instances of both entities represent the same real or abstract thing. One entity (generic entity) represents the complete set of things, the other (category entity) represents a sub-type or sub-classification of those things. The category entity may have one or more characteristics, or a relationship with instances of another entity, not shared by all generic entity instances. Each instance of the category entity is simultaneously an instance of the generic entity.
- Non-specific relationships: A relationship in which an instance of either entity can be related to any number of instances of the other.
- View levels: Three levels of view are defined in IDEF1X: entity relationship (ER), key-based (KB), and fully attributed (FA). They differ in level of abstraction. The ER level is the most abstract. It models the most fundamental elements of the subject area - the entities and their relationships. It is usually broader in scope than the other levels. The KB level adds keys and the FA level adds all the attributes.

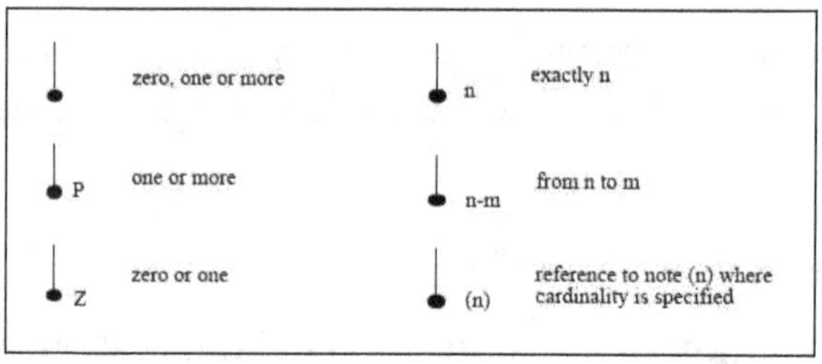

Relationship cardinality syntax.

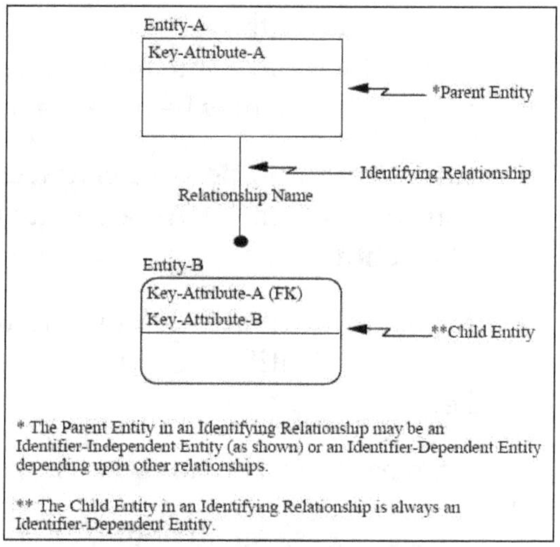

Identifying relationship syntax.

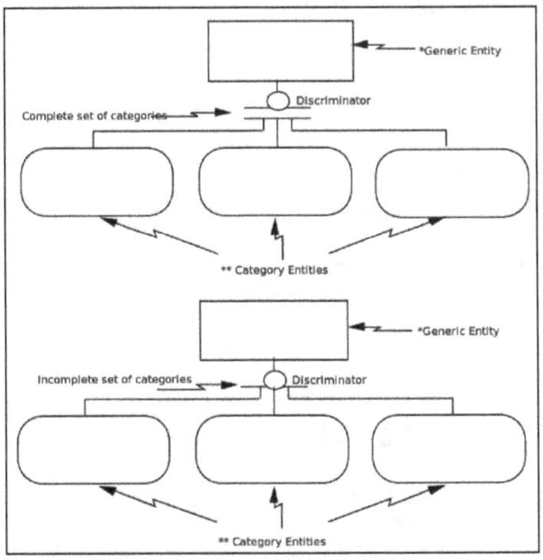

Categorization relationship syntax.

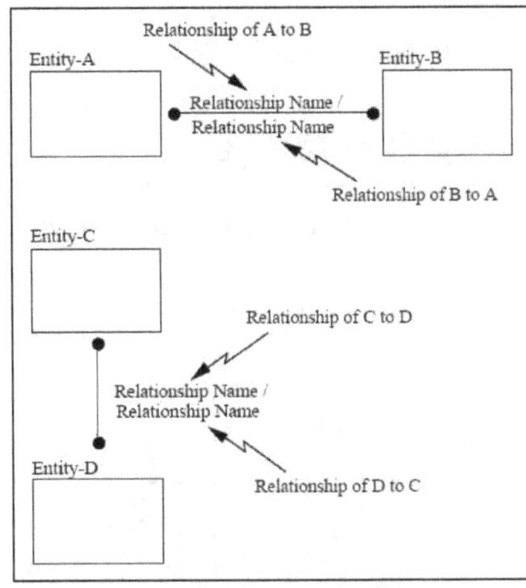

Non-specific relationship syntax.

The Three Schema Approach

The three-schema approach in software engineering is an approach to building information systems and systems information management, that promotes the conceptual model as the key to achieving data integration.

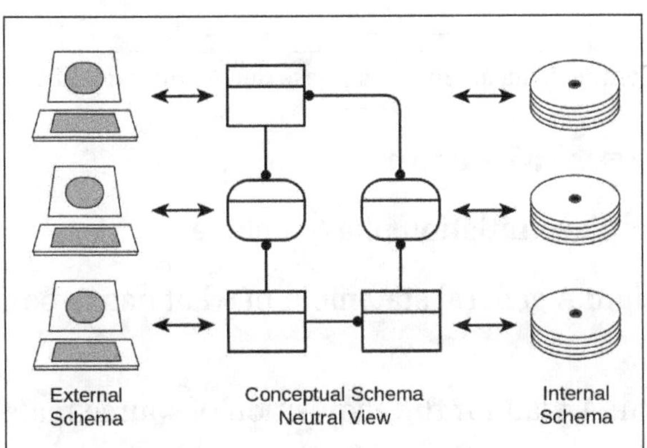

The three schema approach.

A schema is a model, usually depicted by a diagram and sometimes accompanied by a language description. The three schemas used in this approach are:

- External schema for user views.
- Conceptual schema integrates external schemata.
- Internal schema that defines physical storage structures.

At the center, the conceptual schema defines the ontology of the concepts as the users think of them and talk about them. The physical schema describes the internal formats of the data stored in the database, and the external schema defines the view of the data presented to the application programs. The framework attempted to permit multiple data models to be used for external schemata.

Modeling Guidelines

The modeling process can be divided into five stages of model developing.

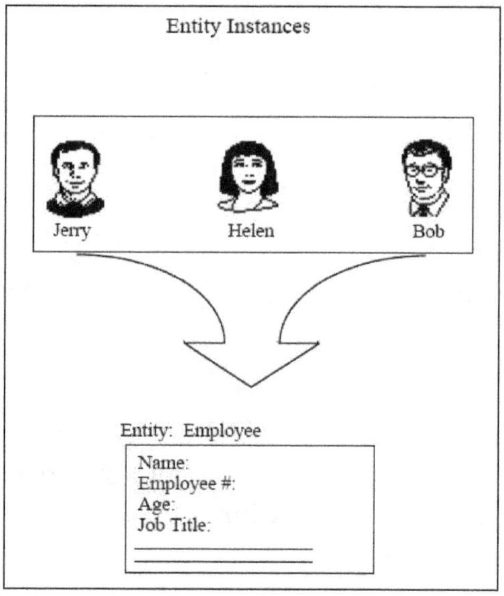

Synthesizing an entity in phase one – entity definition.

Phase Zero – Project Initiation

The objectives of the project initiation phase include:

- Project definition: A general statement of what has to be done, why, and how it will get done.
- Source material: A plan for the acquisition of source material, including indexing and filing.
- Author conventions: A fundamental declaration of the conventions (optional methods) by which the author chooses to make and manage the model.

Phase One – Entity Definition

The objective of the entity definition phase is to identify and define the entities that fall within the problem domain being modeled.

Phase Two – Relationship Definition

The objective of the relationship definition phase is to identify and define the basic relationships between entities. At this stage of modeling, some relationships may be non-specific and will require additional refinement in subsequent phases. The primary outputs from phase two are:

- Relationship matrix,
- Relationship definitions,
- Entity-level diagrams.

Phase Three - Key Definitions

The objectives of the key definitions phase are to:

- Refine the non-specific relationships from phase two.
- Define key attributes for each entity.
- Migrate primary keys to establish foreign keys.
- Validate relationships and keys.

Phase Four - Attribute Definition

The objectives of the attribute definition phase are to:

- Develop an attribute pool,
- Establish attribute ownership,
- Define monkey attributes,
- Validate and refine the data structure.

IDEF1X Meta Model

A meta model is a model of the constructs of a modeling system. Like any model, it is used to represent and reason about the subject of the model - in this case IDEF1X. The meta model is used to reason about IDEF1X, i.e., what the constructs of IDEF1X are and how they relate to one another. The model shown is an IDEF1X model of IDEF1X. Such meta models can be used for various purposes, such as repository design, tool design, or in order to specify the set of valid IDEF1X models. Depending on the purpose, somewhat different models result. There is no "one right model." For example, a model for a tool that supports building models incrementally must allow incomplete or even inconsistent models. The meta model for formalization, however, emphasizes alignment with the concepts of the formalization and hence incomplete or inconsistent models are not allowed.

Meta models have two important limitations. First, they specify syntax but not semantics. Second, a meta model must be supplemented with constraints in natural or formal language. The formal theory of IDEF1X provides both the semantics and a means to precisely express the necessary constraints.

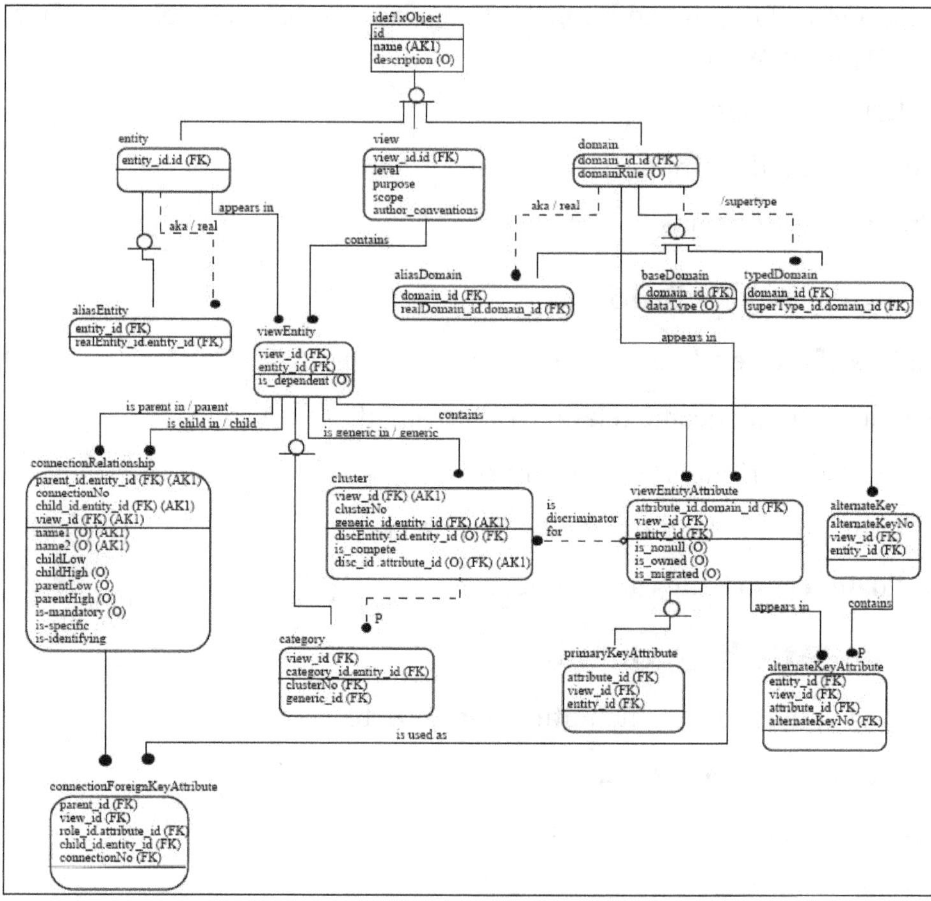

Meta model of IDEF1X.

A meta model for IDEF1X is given in the adjacent figure. The name of the view is *mm*. The domain hierarchy and constraints are also given. The constraints are expressed as sentences in the formal theory of the meta model. The meta model informally defines the set of valid IDEF1X models in the usual way, as the sample instance tables that correspond to a valid IDEF1X model. The meta model also formally defines the set of valid IDEF1X models in the following way. The meta model, as an IDEF1X model, has a corresponding formal theory. The semantics of the theory are defined in the standard way. That is, an interpretation of a theory consists of a domain of individuals and a set of assignments:

- To each constant in the theory, an individual in the domain is assigned.
- To each n-ary function symbol in the theory, an n-ary function over the domain is assigned.
- To each n-ary predicate symbol in the theory, an n-ary relation over the domain is assigned.

In the intended interpretation, the domain of individuals consists of views, such as production; entities, such as part and vendor; domains, such as qty_on_hand; connection relationships; category clusters; and so on. If every axiom in the theory is true in the interpretation, then the interpretation is called a model for the theory. Every model for the IDEF1X theory corresponding to the IDEF1X meta model and its constraints is a valid IDEF1X model.

IDEF3

IDEF3 or Integrated DEFinition for Process Description Capture Method is a business process modelling method complementary to IDEF0. The IDEF3 method is a scenario-driven process flow description capture method intended to capture the knowledge about how a particular system works.

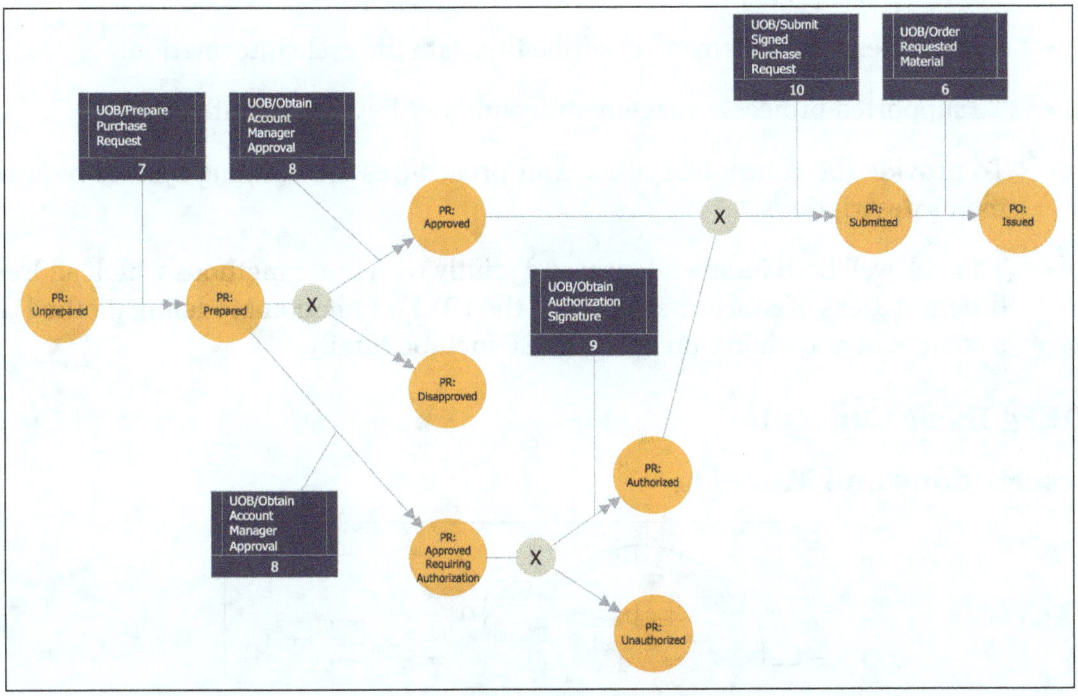

Example of an Enhanced Transition Schematic, modelled with IDEF3.

The IDEF3 method provides modes to represent both:

- Process Flow Descriptions to capture the relationships between actions within the context of a specific scenario.
- Object State Transition to capture the description of the allowable states and conditions.

This method is part of the IDEF family of modeling languages in the field of systems and software engineering.

One of the primary mechanisms used for descriptions of the world is relating a story in terms of an ordered sequence of events or activities. The IDEF3 Process Description

Capture Method was created to capture descriptions of sequences of activities, which is considered the common mechanisms to describe a situation or process. The primary goal of IDEF3 is to provide a structured method by which a domain expert can express knowledge about the operation of a particular system or organization. Knowledge acquisition is enabled by direct capture of assertions about real-world processes and events in a form that is most natural for capture. IDEF3 supports this kind of knowledge acquisition by providing a reliable and wellstructured approach for process knowledge acquisition, and an expressively, yet easy-to-use, language for information capture and expression.

Motives for the development of IDEF3 were the need:

- To speed up the process of business systems modeling.
- To provides mechanisms to describe this data life cycle information.
- To supported project management techniques by an automated tool.
- To provide the concepts, syntax, and procedures for building system requirements descriptions.
- To work well both independently and jointly with other methods which address different areas of concentration (e.g., the IDEF0 Function Modeling method) as a complementary addition to the IDEF method family.

IDEF3 Basic Concepts

Descriptions and Models

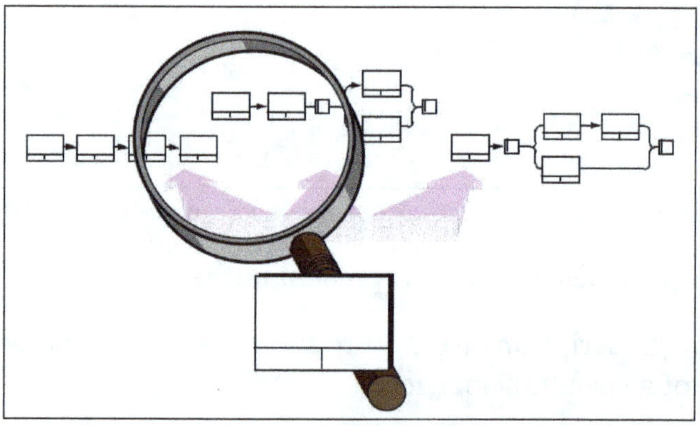

IDEF3 want to offer alternative descriptions of the same process from multiple viewpoints on the process.

The distinction between descriptions and models, though subtle, is an important one in IDEF3, and both have a precise technical meaning.

- The term description is used as a reserved technical term to mean records of empirical observations; that is, descriptions record knowledge that originates in or is based on observations or experience.

- The term model is used to mean an idealization of an entity or state of affairs. That is, a model constitutes an idealized system of objects, properties, and relations that is designed to imitate, in certain relevant respects, the character of a given real-world system. Frictionless planes, perfectly rigid bodies, the assumption of point mass, and so forth are representative examples of models.

The power of a model comes from its ability to simplify the real-world system it represents and to predict certain facts about that system by virtue of corresponding facts within the model. Thus, a model is a designed system in its own right. Models are idealized systems known to be incorrect but assumed to be close enough to provide reliable predictors for the predefined areas of interest within a domain. A description, on the other hand, is a recording of facts or beliefs about something within the realm of an individual's knowledge or experience. Such descriptions are generally incomplete; that is, the person giving a description may omit facts that he or she believes are irrelevant, or which were forgotten in the course of describing the system. Descriptions may also be inconsistent with respect to how others have observed situations within the domain. IDEF3 accommodates these possibilities by providing specific features enabling the capture and organization of alternative descriptions of the same scenario or process.

Description Capture

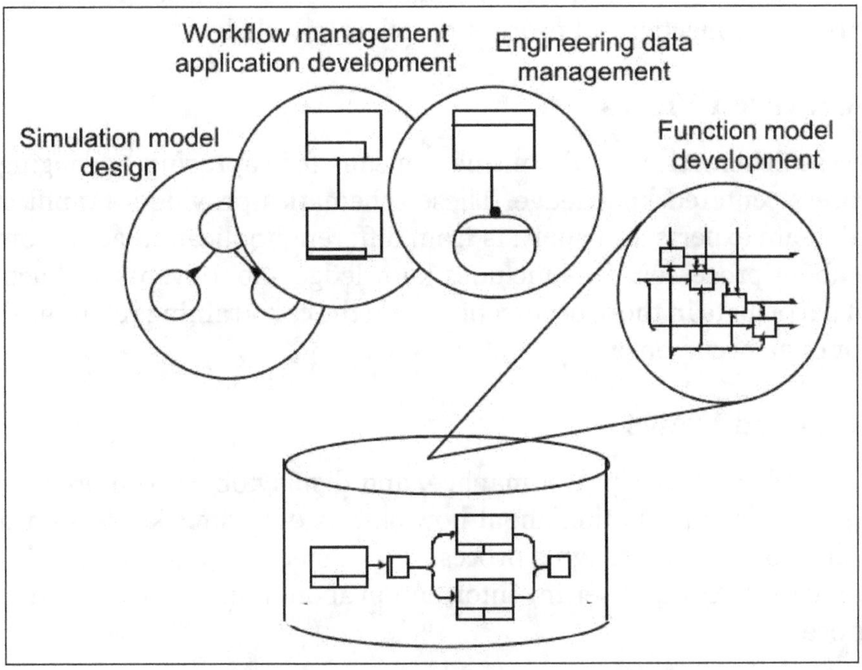

IDEF focus on Description Capture enabling reuse.

Modeling necessitates taking additional steps beyond description capture to resolve conflicting or inconsistent views. This, in turn, generally requires modelers to select or create a single viewpoint and introduce artificial modeling approximations to fill in gaps where no direct knowledge or experience is available. Unlike models, descriptions are not constrained by idealized, testable conditions that must be satisfied, short of simple accuracy.

The purpose of description capture may be simply to record and communicate process knowledge or to identify inconsistencies in the way people understand how key processes actually operate. By using a description capture method users need not learn and apply conventions forcing them to produce executable models (e.g., conventions ensuring accuracy, internal consistency, logical coherence, non-redundancy, completeness). Forcing users to model requires them to adopt a model design perspective and risk producing models that do not accurately capture their empirical knowledge of the domain.

Scenarios

The notion of a scenario or story is used as the basic organizing structure for IDEF3 Process Descriptions. A scenario can be thought of as a recurring situation, a set of situations that describe a typical class of problems addressed by an organization or system, or the setting within which a process occurs. Scenarios establish the focus and boundary conditions of a description. Using scenarios in this way exploits the tendency of humans to describe what they know in terms of an ordered sequence of activities within the context of a given scenario or situation. Scenarios also provide a convenient vehicle to organize collections of process-centered knowledge.

Process-centered Views

IDEF3 Process Schematics are the primary means for capturing, managing, and displaying process-centered knowledge. These schematics provide a graphical medium that helps domain experts and analysts from different application areas communicate knowledge about processes. This includes knowledge about events and activities, the objects that participate in those occurrences, and the constraining relations that govern the behavior of an occurrence.

Object-centered Views

IDEF3 Object Schematics capture, manage, and display object-centered descriptions of a process—that is, information about how objects of various kinds are transformed into other kinds of things through a process, how objects of a given kind change states through a process, or context-setting information about important relations among objects in a process.

IDEF3 Process Description Language

IDEF3 descriptions are developed from two different perspectives: process-centered and object-centered. Because these approaches are not mutually exclusive, IDEF3 allows cross-referencing between them to represent complex process descriptions.

Process Schematics

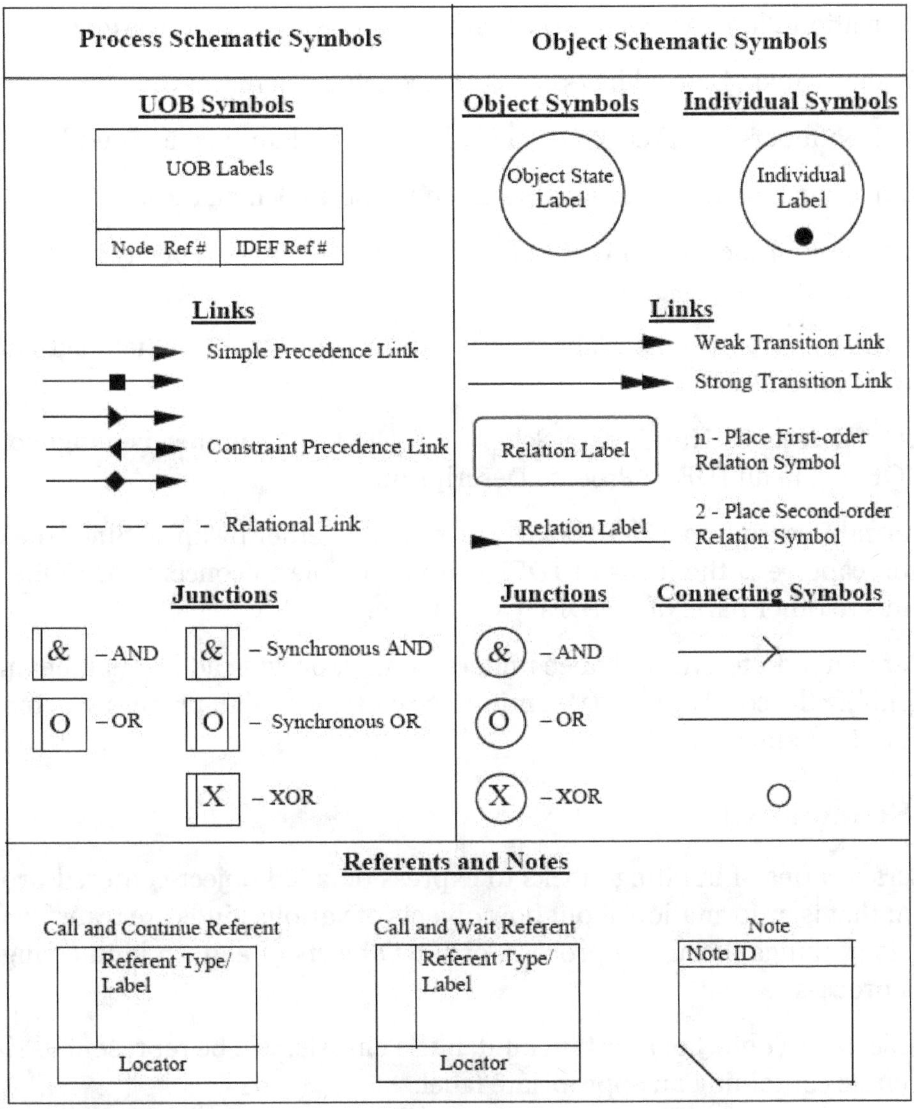

Symbols Used for IDEF3 Process Description Schematics.

Process schematics tend to be the most familiar and broadly used component of the IDEF3 method. These schematics provide a visualization mechanism for process-centered descriptions of a scenario. The graphical elements that comprise process schematics include Unit of Behavior (UOB) boxes, precedence links, junctions, referents, and notes. The building blocks here are:

- Unit of Behavior (UOB) boxes.

- Links: Links are the glue that connect UOB boxes to form representations of dynamic processes.
- Simple Precedence Links: Precedence links express temporal precedence relations. between instances of one UOB and those of another.
- Activation Plots: Activation plots are used to represent activations.
- Dashed Links: Dashed links carry no predefined semantics.
- Link Numbers: All links have an elaboration and unique link numbers.
- Activation Semantics for Nonbranching Process Schematics.
- Junctions: Junctions in IDEF3 provide a mechanism to specify the logic of process branching.
- UOB Decompositions: Elaborations capture and structure detailed knowledge about processes.
- UOB Reference Numbering Scheme: A UOB box number is assigned to each UOB box in an IDEF3 Process Description.
- Partial Descriptions: UOB boxes are joined together by links. Since the description capture is the focus of IDEF3, it is possible to conceive of UOBs without links to other parts of an IDEF3 schematic.
- Referents: Referents enhance understanding, provide additional meaning, and simplify the construction (i.e., minimize clutter) of both process schematics and object schematics.

Object Schematics

IDEF offers a series of building blocks to express detailed object-centered process information; that is, information about how objects of various kinds are transformed into other kinds of things through a process, or how objects of a given kind change states through a process.

- Objects: An object of a certain kind, like a chassis, will be represented simply by a circle containing an appropriate label.
- Object States: A certain kind of object being in a certain state will be represented by a circle with a label that captures the kind itself and a corresponding state, representing thereby the type, or class, of objects that are in that state.
- Object schematics: The construction of complex representations built from kind symbols and object state symbols.
- Transition Schematics: The first and most basic construct is the basic state transition schematic or simply, transition schematic.

IDEF4

IDEF4, or Integrated DEFinition for Object-Oriented Design, is an object-oriented design modeling language for the design of component-based client/server systems. It has been designed to support smooth transition from the application domain and requirements analysis models to the design and to actual source code generation. It specifies design objects with sufficient detail to enable source code generation.

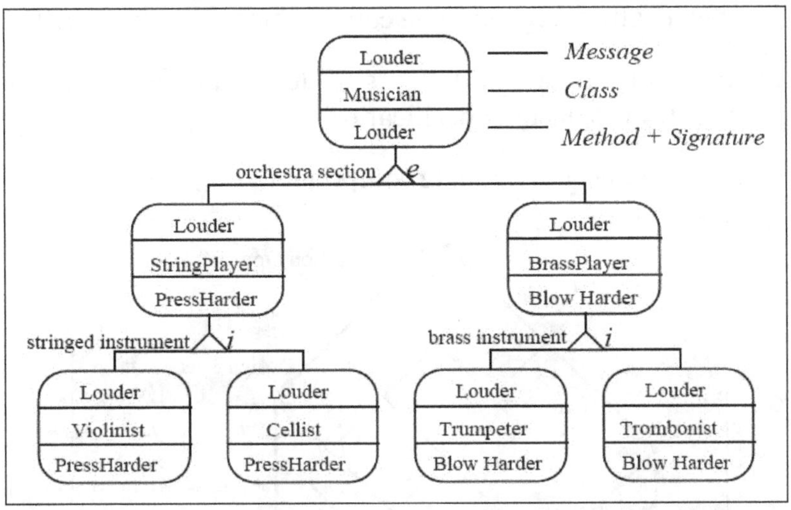

Example of the IDEF4: An Behavior Diagram for methods Implementing Louder.

This method is part of the IDEF family of modeling languages in the field of systems and software engineering.

IDEF4 method is a graphically oriented methodology for the design of object-oriented software systems. The object-oriented programming paradigm provides the developer with an abstract view of his program as composed of a set of state maintaining objects which define the behavior of the program by the protocol of their interactions. An object consists of a set of local state defining attributes and a set of methods (procedures) that define the behavior of that particular object and its relationship to the other objects that make up the system.

The IDEF4 method multi-dimensional approach to object-oriented software system design consists of the following items:

- Design layers (system-level, application-level, and low-level design).
- Artifact design status (application domain, in transition, software domain).
- Design models (static, dynamic, and behavior) and the design rationale component.
- Design features ranging from general to specific enabling deferred decision making.

IDEF4 Concepts

Dimensions of IDEF4 Design Objects

IDEF4 uses an object-oriented design method or procedure that is very similar to Rumbaugh's Object Method Technique and Schlaer/Mellor's Object-Oriented Analysis and Design (OOA/OOD) technique. However, there are some crucial differences:

- IDEF4 is specifically designed to be compatible with other IDEF methods.
- IDEF4 allows one to track the status of design artifacts from domain object through transition to design specification.
- IDEF4 includes a design rationale component.

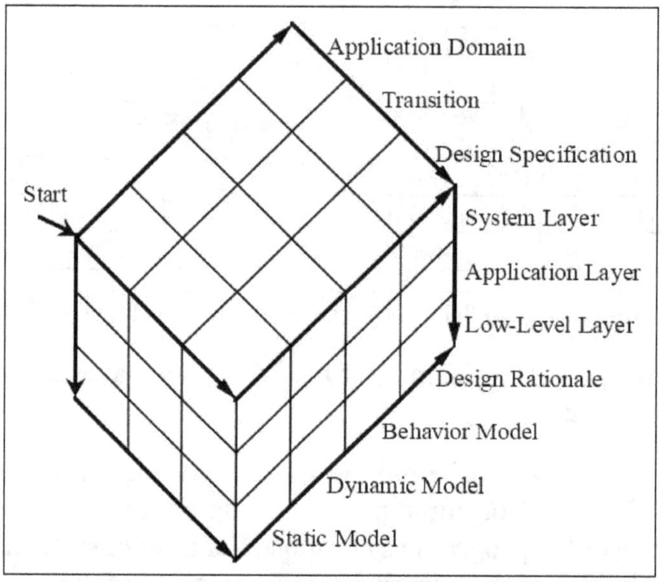

Dimensions of IDEF4 Design Objects.

These extra dimensions are shown in the figure. The edges of the box show the progression of the design from start to finish elaborating each of these dimensions.

IDEF4 Design Activities

In IDEF4, a design starts with the analysis of requirements and takes as input the domain objects. These domain objects are encoded in their equivalent IDEF4 form and marked as domain objects. As computational objects are developed for these objects, they are marked as "transitional" and finally as "completed." The level of completion of an IDEF4 design is determined by setting measures based on the status, level, and model dimensions of individual artifacts in the design.

The system-level design starts once the "raw material" (domain) objects have been collected. This develops the design context, ensures connectivity to legacy systems, and identifies the applications that must be built to satisfy the requirements. Static, dynamic, behavioral, and rationale models are built for the objects at the system level. These specifications become the requirements on the application level – the next level of design. The application level design identifies and specifies all of the software components (partitions) needed in the design. Static models, dynamic models, behavioral models, and the rationale component are built for the objects at the application level. These specifications become the requirements on the next level of design – the low-level design. Static Models, Dynamic Models, Behavioral Models, and the design rationale component are built for the low-level design objects. Sub-layers may be built within each layer to reduce complexity.

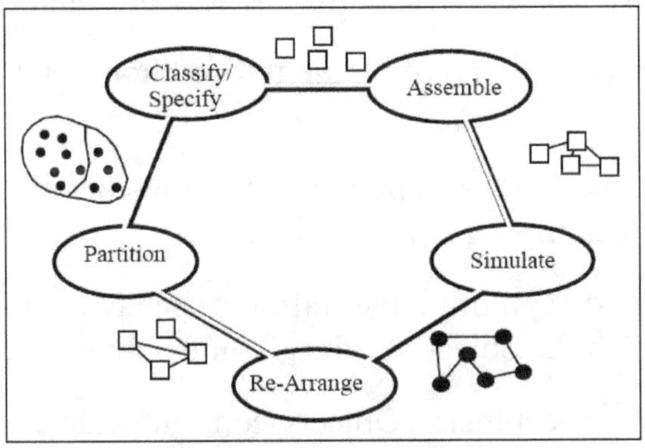

IDEF4 Design Activities.

IDEF4 is an iterative procedure involving partitioning, classification/specification, assembly, simulation, and re-partitioning activities. First the design is partitioned into objects, each of which is either classified against existing objects or for which an external specification is developed. The external specification enables the internal specification of the object to be delegated and performed concurrently. After classification/specification, the interfaces between the objects are specified in the assembly activity (i.e., static, dynamic, and behavioral models detailing different aspects of the interaction between objects are developed). While the models are developed, it is important to simulate use scenarios or cases between objects to uncover design flaws. Based on these flaws the designer can then rearrange the existing models and simulate them until the designer is satisfied.

IDEF4 Object-oriented Concepts

IDEF4's defines a set of object oriented concepts:

- Domains: IDEF4 projects are implemented in a domain. A domain can be seen as the scope of the system being developed. During system design, the software is transitioned between three domains: the application domain, the design domain, and the implementation domain.
- Features, Artifacts, and Objects.

- Object Instance: Objects can be object instances, object classes, and object partitions. Object instances are the individual things encountered in the application domain.
- Classes: Classes are generalizations about objects and are used to manage complexity by taking advantage of similarities in object instances and grouping them under a class or category.
- Subclass/Superclass: The term subclass captures the concept of grouping particular instances of a class into an even more specialized class.
- Partitions: A partition object contains objects and relations.
- Attributes: Attributes are an implementation choice on how to represent an object's state.
- Object States: Object states represent situations or conditions of an object instance that are meaningful in the design.
- Method: A method is an implementation of behavior (i.e., a set of instructions according to which the object performs some operation).
- Message and Polymorphism: Objects communicate by sending messages to each other.
- Event: An event is a signal generated by a method in an object indicating some condition in the object.
- Object Life cycles: In any system, objects exhibit patterns of behavior as they cycle through different states.
- Client/Server: An object plays the role of a client relative to a message if it is the sender of that message.
- Relationships and Roles: Objects connected together with arcs. These arcs are called relationships and they show associations between objects.
- Inheritance: A specific type of relationship used in object-oriented technology is inheritance.
- Encapsulation and Information Hiding: Encapsulation and information hiding are two object-oriented concepts that are most easily understood when discussed in terms of interactions between objects.

Object Class Identification

The IDEF4 Method assumes that the domain objects have been identified through Object-Oriented Domain Analysis. Methods such as IDEF1, IDEF5, IDEF3, SA/SD can be used to perform domain analysis. However, IDEF4 practitioners should be aware of how objects are identified, as the design process may reveal deficiencies in the Object-Oriented Analysis. IDEF4 had defined five types of classes:

- Physical Objects.
- Role Objects: The role may be related to other activities that the person engages in (e.g., a patient in a hospital, a stockholder, a client, a trustee, a suspect in a burglary, or a tax payer).
- Event Objects: Events or incidents may also be considered objects. The identification of events as objects is highly subjective, and will depend on the domain in which the software is to be used.
- Interaction Objects: Interaction objects are the result of interactions or transactions between two or more objects.
- Specification and Procedure Objects: Specification objects describe the acceptable characteristics of objects instances. Procedure objects refer to the way other object instances may interact.

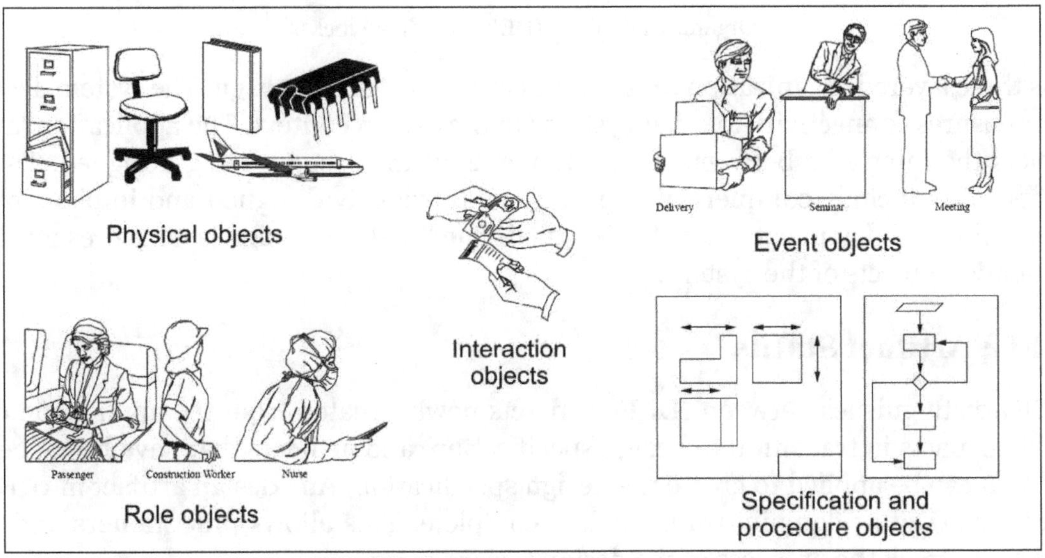

The five types of Object classes in IDEF4.

IDEF4 Building Blocks

IDEF4 Layers

IDEF4 users design in three distinct layers:

- System design.
- Application design.

- Low-level design.

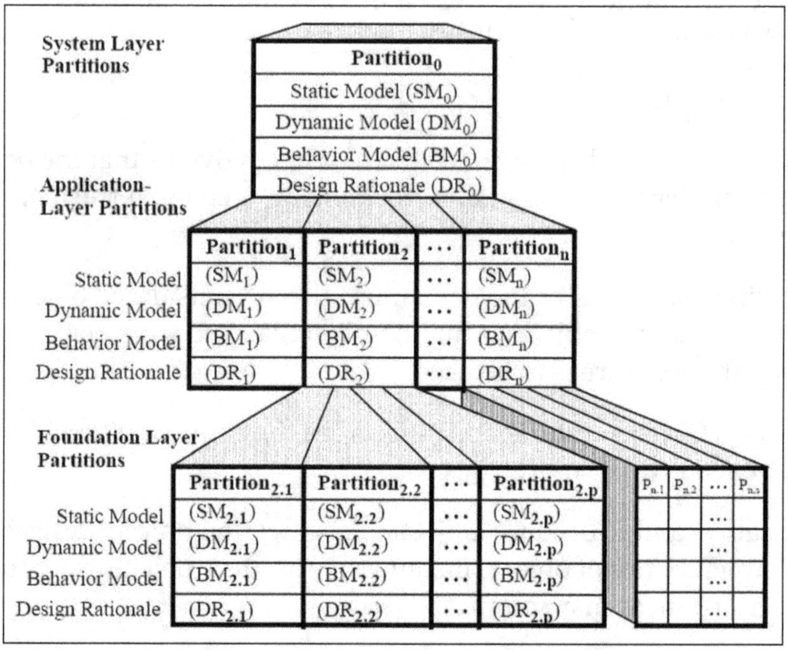

Organization of the IDEF4 Building blocks.

This three layered organization reduces the complexity of the design. The system design layer ensures connectivity to other systems in the design context. The application layer depicts the interfaces between the components of the system being designed. These components include commercial applications, previously designed and implemented applications, and applications to be designed. The low-level design layer represents the foundation objects of the system.

IDEF4 Artifact Status

IDEF4 distinguishes between IDEF4 artifacts newly created from the application domain, artifacts in transition to design specification, and artifacts that have been specified that can be applied to create the design specification. Any design artifact in IDEF4 can be marked as domain, transition, or complete. This allows practitioners and reviewers to track the progress of the design toward completion.

IDEF4 Design Models

IDEF4 uses three design models and a design rationale component:

- The Static Model (SM) defines time-invariant relations between objects (for example, inheritance).
- The Dynamic Model (DM) specifies the communication between objects and the state transitions of objects.

- The Behavior Model (BM) defines the relations between the respective behaviors of objects.

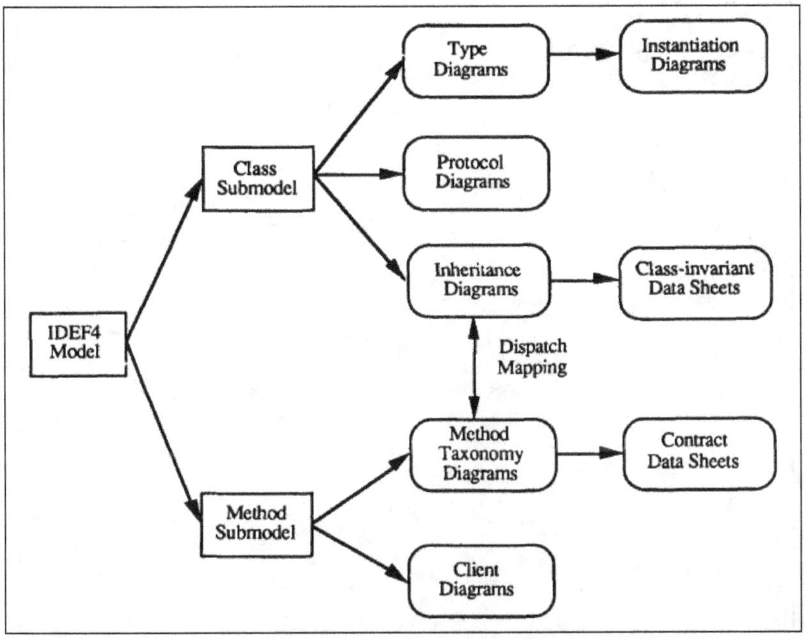

Organization of the IDEF4 model.

The design rationale component provides a top-down representation of the system, giving a broad view that encompasses the three design models and documents the rationale for major design evolutions.

Each model represents a different cross section of the design. The three design models capture all the information represented in a design project, and the design rationale documents the reasoning behind the design. Each model is supported by a graphical syntax that highlights the design decisions that must be made and their impact on other perspectives of the design. To facilitate use, the graphical syntax is identical among the three models.

Design Features

IDEF4 provides a broad range of design features – from generic to specific. This range enables deferred decision making by allowing the designer to first capture design features in general terms and later to refine them. This significantly reduces the burden on designers by allowing them to immediately capture new design concepts with IDEF4 design features.

IDEF5

IDEF5 (Integrated Definition for Ontology Description Capture Method) is a software engineering method to develop and maintain usable, accurate domain ontologies. This standard is part of the IDEF family of modeling languages in the field of software engineering.

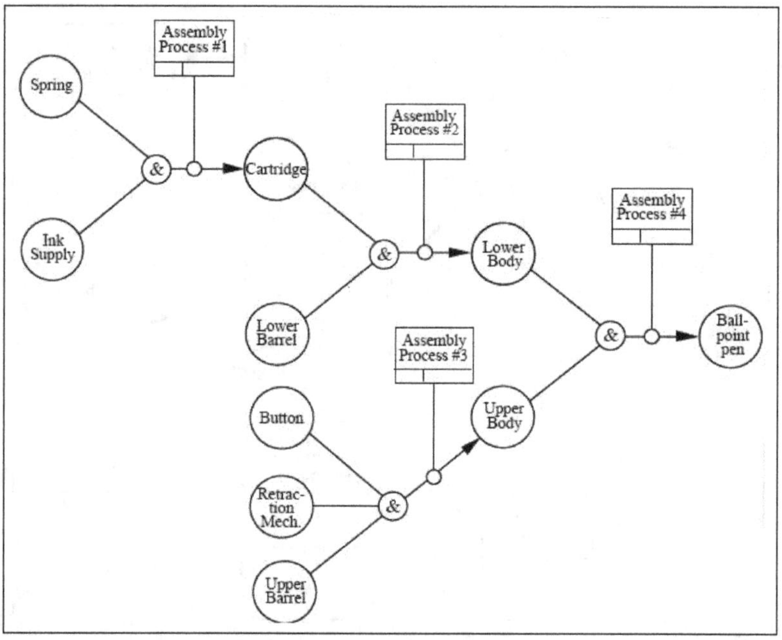

Example of an IDEF5 Composition Schematic for a Ballpoint Pen.

In the field of computer science ontologies are used to capture the concept and objects in a specific domain, along with associated relationships and meanings. In addition, ontology capture helps coordinate projects by standardizing terminology and creates opportunities for information reuse. The IDEF5 Ontology Capture Method has been developed to reliably construct ontologies in a way that closely reflects human understanding of the specific domain.

In the IDEF5 method, an ontology is constructed by capturing the content of certain assertions about real-world objects, their properties, and their interrelationships and representing that content in an intuitive and natural form. The IDEF5 method has three main components:

- A graphical language to support conceptual ontology analysis.
- A structured text language for detailed ontology characterization.
- A systematic procedure that provides guidelines for effective ontology capture.

Topics

Ontology

In IDEF5 the meaning of the term ontology is characterized to include a catalog of terms used in a domain, the rules governing how those terms can be combined to make valid statements about situations in that domain, and the "sanctioned inferences" that can be made when such statements are used in that domain. In every domain, there are phenomena that the humans in that domain discriminate as (conceptual or physical) objects, associations, and situations. Through various language mechanisms, one associates definite descriptors (e.g., names, noun phrases, etc.) to those phenomena.

Central Concepts of Ontology

The construction of ontologies for human engineered systems is the focus of the IDEF5. In the context of such systems, the nature of ontological knowledge involves several modifications to the more traditional conception. The first of these modifications has to do with the notion of a kind. Historically, a kind is an objective category of objects that are bound together by a common nature, a set of properties shared by all and only the members of the kind.

While there is an attempt to divide the world at its joints in the construction of enterprise ontologies, those divisions are not determined by the natures of things in the enterprise so much as the roles those things are to play in the enterprise from some perspective or other. Because those roles might be filled in any of a number of ways by objects that differ in various ways, and because legitimate perspectives on a domain can vary widely, it is too restrictive to require that the instances of each identifiable kind in an enterprise share a common nature, let alone that the properties constituting that nature be essential to their bearers. Consequently, enterprise ontologies require a more flexible notion of kind.

Ontology Development Process

Ontology development requires extensive iterations, discussions, reviews, and introspection. Knowledge extraction is usually a discovery process and requires considerable introspection. It requires a process that incorporates both significant expert involvement as well as the dynamics of a group effort. Given the open-ended nature of ontological analyses, it is not prudent to adopt a "cookbook" approach to ontology development. In brief, the IDEF5 ontology development process consists of the following five activities:

- Organizing and Scoping: This activity involves establishing the purpose, viewpoint, and context for the ontology development project and assigning roles to the team members.
- Data Collection: This activity involves acquiring the raw data needed for ontology development.
- Data Analysis: This activity involves analyzing the data to facilitate ontology extraction.
- Initial Ontology Development: This activity involves developing a preliminary ontology from the acquired data.

- Ontology Refinement and Validation: This activity involves refining and validating the ontology to complete the development process.

Although the above activities are listed sequentially, there is a significant amount of overlap and iteration between the activities.

Ontological Analysis

Ontological analysis is accomplished by examining the vocabulary that is used to discuss the characteristic objects and processes that compose the domain, developing rigorous definitions of the basic terms in that vocabulary, and characterizing the logical connections among those terms. The product of this analysis, an ontology, is a domain vocabulary complete with a set of precise definitions, or axioms, that constrain the meanings of the terms sufficiently to enable consistent interpretation of the data that use that vocabulary.

IDEF5 Building Blocks

Some of the key terms in IDEF5 and the basic IDEF5 Schematic Language Symbols:

- Kind: Informally, a group of individuals that share some set of distinguished characteristics. More formally, kinds are properties typically expressed by common nouns such as 'employee', 'machine', and 'lathe'.
- Individual: The most logically basic kind of real world object. Prominent examples include human persons, concrete physical objects, and certain abstract objects such as programs. Unlike objects of higher logical orders such as properties and relations, individuals essentially are not multiply instantiable. Individuals are also known as first-order objects.
- Referent: A construct in the IDEF5 elaboration language used to refer to a kind, object, property, relation, or process kind in another ontology or an IDEF model.
- Relation: An abstract, general association or connection that holds between two or more objects. Like properties, relations are multiply instantiable. The objects among which a relation holds in a particular instance are known as its arguments.
- State: A property, generally indicated by an adjective rather than a common noun, that is characteristic of objects of a certain kind at a certain point within a process. For example, water can be in frozen, liquid, or gaseous states.
- Process: A real world event or state of affairs involving one or more individuals over some (possibly instantaneous) interval of time. Typically, a process involves some sort of change in the properties of one or more of the individuals

within the process. Because of the ambiguity in the term "process", sometimes referred to as process instance.

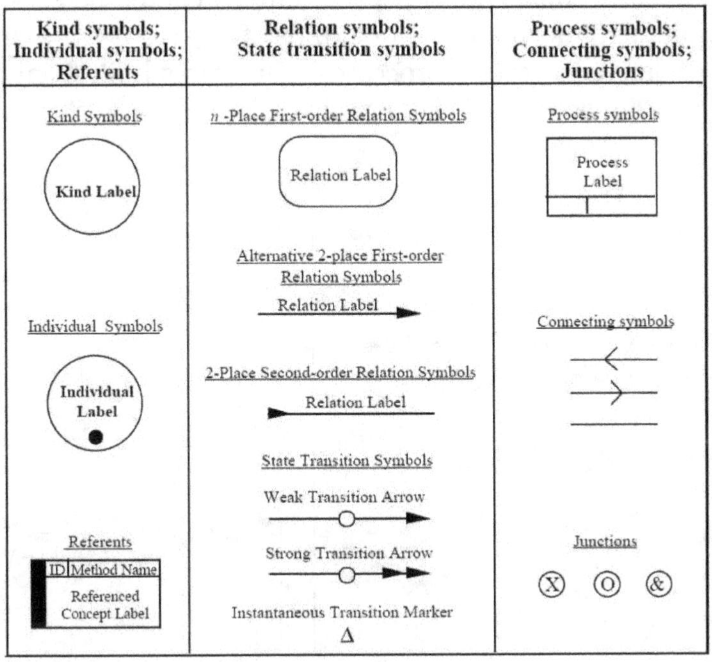

Diagram Types

Various diagram types, or schematics, can be constructed in the IDEF5 Schematic Language. The purpose of these schematics, like that of any representation, is to represent information visually. Thus, semantic rules must be provided for interpreting every possible schematic. These rules are provided by outlining the rules for interpreting the most basic constructs of the language, then applying them recursively to more complex constructs. There are four primary schematic types derived from the basic IDEF5 Schematic Language which can be used to capture ontology information directly in a form that is intuitive to the domain expert.

- Classification Schematics: Classification schematics provide mechanisms for humans to organize knowledge into logical taxonomies. Of particular merit are two types of classification: description subsumption and natural kind classification.

- Composition Schematics: Composition schematics serve as mechanisms to represent graphically the "part-of" relation that is so common among components of an ontology.

- Relation Schematics: Relation schematics allow ontology developers to visualize and understand relations among kinds in a domain, and can also be used to capture and display relations between first-order relations.

- Object State Schematics: Because there is no clean division between information about kinds and states and information about processes, the IDEF5 schematic language enables modelers to express fairly detailed object-centered process information (i.e., information about kinds of objects and the various states they can be in relative to certain processes). Diagrams built from these constructs are known as Object-State Schematics.

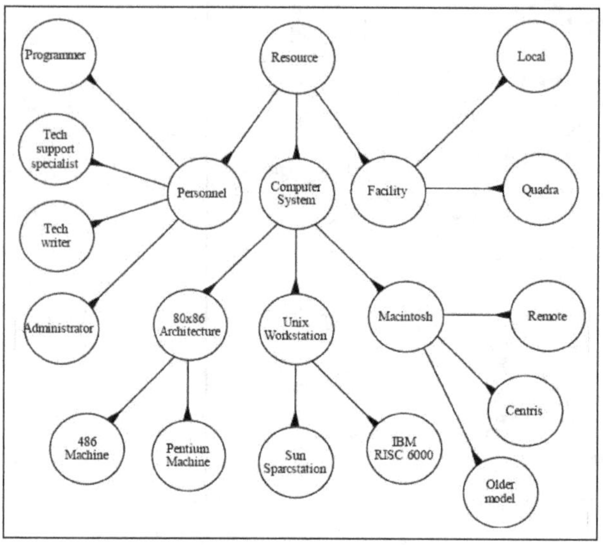

Classification Schematics.

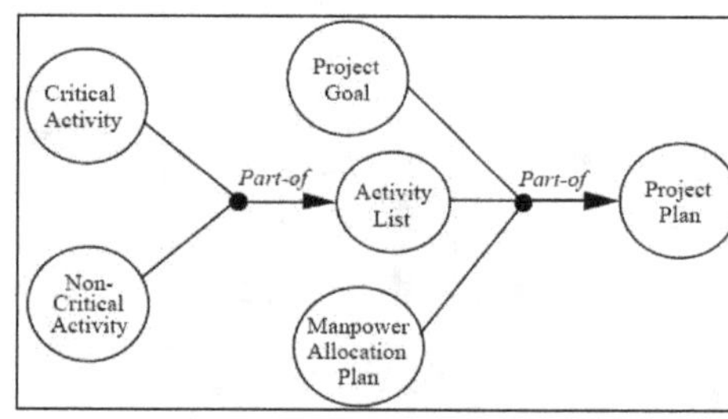

Composition Schematics.

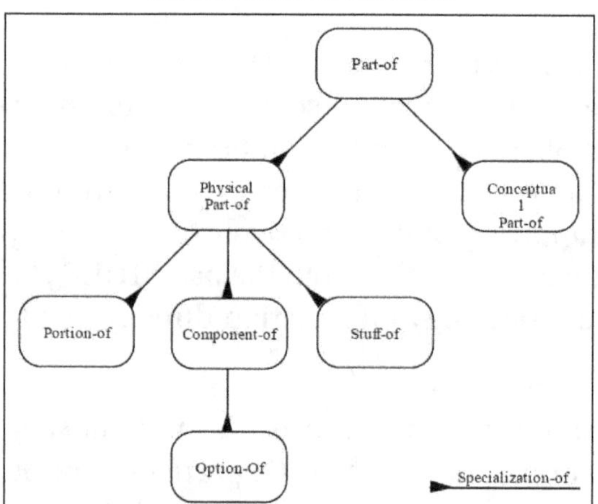

Relation Schematics.

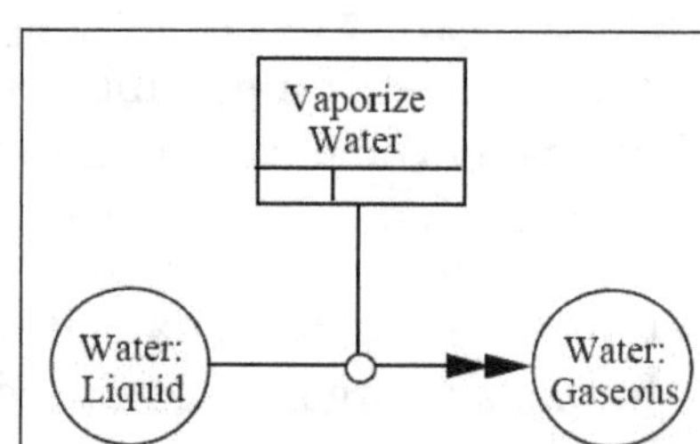

Object State Schematics.

IDEF6

IDEF6 or Integrated Definition for Design Rationale Capture is a method to facilitate the acquisition, representation, and manipulation of the design rationale used in the development of enterprise systems. This method, that wants to define the motives that drive the decision-making process, is still in development. Rationale is the reason, justification, underlying motivation, or excuse that moved the designer to select a particular strategy or design feature. More simply, rationale is interpreted as the answer to the question, "Why is this design being done in this manner?" Most design methods focus on what the design is (i.e., on the final product, rather than why the design is the way it is).

IDEF6 is part of the IDEF family of modeling languages in the field of systems and software engineering.

When explicitly captured, design rationale typically exists in the form of unstructured textual comments. In addition to making it difficult, if not impossible to find relevant information on demand, lack of a structured method for organizing and providing completeness criteria for design rationale capture makes it unlikely that important information will be documented.

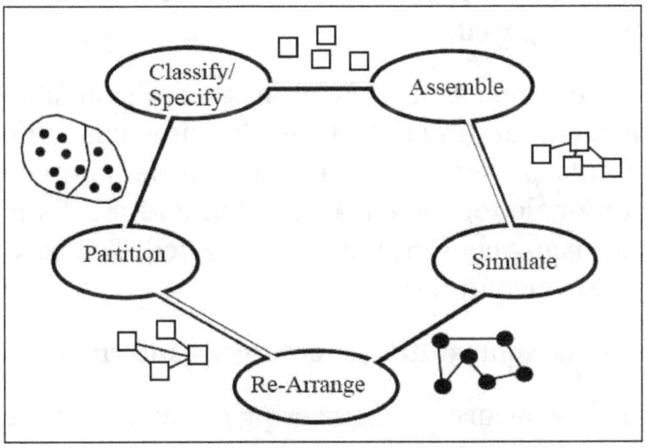

IDEF6 model of IDEF4 Design Activities.

IDEF6 was intended to be a method with the representational capability to capture information system design rationale and associate that rationale with the design models and documentation for the end system. Thus, IDEF6 attempts to capture the logic underlying the decisions contributing to, or resulting in, the final design. The explicit capture of design rationale serves to help avoid repeating past mistakes, provides a direct means for determining the impact of proposed design changes, forces the explicit statement of goals and assumptions, and aids in the communication of final system specifications.

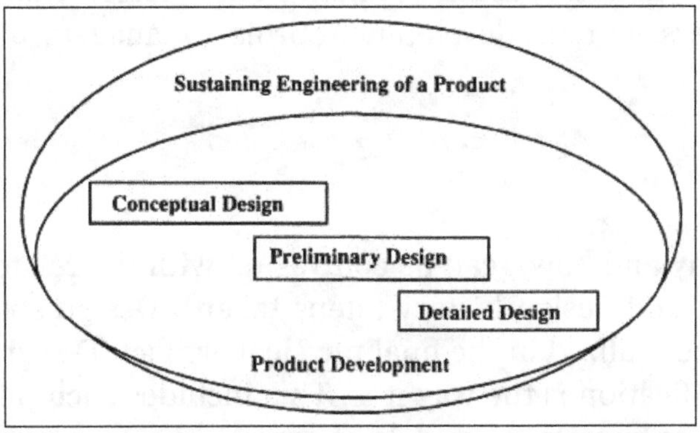

The predominate modes of design commonly found in information system development.

IDEF6 will be a method that possesses the conceptual resources and linguistic capabilities needed.

- To represent the nature and structure of the information that constitutes design rationale within a given system.
- To associate that rationale with design specifications, models, and documentation for the system.

The scope of IDEF6 applicability covers all phases of the information system development process, from initial conceptualization through both preliminary and detailed design activities. To the extent that detailed design decisions for software systems are relegated to the coding phase, the IDEF6 technique should be usable during the software construction process as well.

Design rationale becomes important when a design decision is not completely determined by the constraints of the situation. Thus, decision points must be identified, the situations and constraints associated with those decision points must be defined, and if options exist, the rationale for the chosen option and for discarding other options (i.e., those design options not chosen) must be recorded. The task of capturing design rationale serves the following purposes:

- Enables evolutionary integration of enterprise information systems.
- Enables the use of concurrent engineering methods in information systems development.
- Supports better integration among life-cycle artifacts.
- Facilitates business reengineering by capturing the rationale behind business case decisions.
- Enables efficient traceability of decisions.

Rationale capture is applicable to all phases of the system development process. The intended users of IDEF6 include business system engineers, information systems designers, software designers, systems development project managers, and programmers.

IDEF6 Topics

Basic Concepts

Design rationale (why and how), can be contrasted with the related notions of design specification (what), and design history (steps taken). Design specifications describe what intent should be realized in the final physical artifact. Design rationale describes why the design specification is the way it is. This includes such information as principles and philosophy of operation, models of correct behavior, and models of how the artifact behaves as it fails. The design process history records the steps that were taken, the plans and expectations that led up to these steps, and the results of each step.

- Design Rationale Phenomena: A general characterization of design rationale can be given as: "The beliefs and facts as well as their organization that the human uses to make (or justify) design commitments and to propagate those commitments."
- Design Rationale Capture Issues: One reason for the loss of rationale is rooted in the long time lag between specification of the software artifact and completion of the artifact. There are also problems with developing a general understanding of what should constitute explicitly captured Design Rationale. That is, a notable difficulty with expressing design rationale is that the concept itself is not uniformly understood. It shares this characteristic with all other forms of "explanation" that Artificial Intelligence (AI) researchers continue to struggle with.

Procedure Developments

In IDEF6, the rationale capture procedure involves partitioning, classification/ specification, assembly, simulation/execution, and re-partitioning activities. The rationale capture procedure normally applied in the simulation/execution activity of the evolving design uses two phases: Phase I describes the problem and Phase II develops a solution strategy.

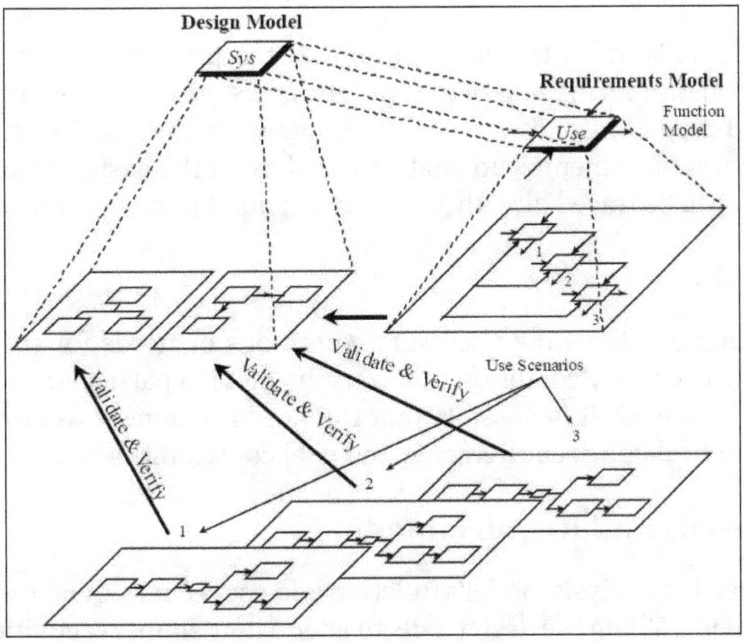

A design partition called Sys, showing its constituent static and dynamic models, as well as its associated requirements model. The requirements model contains an IDEF0 function model whose activities call out IDEF3 process scenarios.

Design is an iterative procedure involving partitioning, classification/specification, assembly, simulation, and re-partitioning activities. First, the design is partitioned into design artifacts. Each artifact is either classified against existing design artifacts or an external specification is developed for it. The external specification enables the internal specification of the design artifact to be delegated and performed concurrently. After classification/specification, the interfaces between the design artifacts are specified in the assembly activity (i.e., static, dynamic, and behavioral models detailing different aspects of the interaction between design artifacts are developed). While the models are developed, it is important to simulate use scenarios or use cases between design artifacts to uncover design flaws. By analyzing these flaws, the designer can re-arrange the existing models and simulate them until the designer is satisfied. The observed design flaws and the actions contemplated and taken for each are the basis of the design rationale capture procedure.

Identify Problems

The designer identifies problems in the current design state by stepping through the use cases in the requirements model to validate that the design satisfies requirements and to verify that the design will function as intended. The designer records symptoms or concerns about the current design state. A symptom is an observation of an operational failure or undesirable condition in the existing design. A concern is an observation of an anticipated failure or undesirable condition in the existing design.

Identify Constraints

The designer then identifies the constraints that the problems violate or potentially violate. These constraints include requirements, goals, physical laws, conventions, assumptions, models, and resources. Because the activities and processes in the use case scenarios map to requirements and goals, the failure of the design in any use case activity or process can be traced directly to requirements statements and goal statements.

Identify Needs

The designer then identifies the necessary conditions or needs for solving the problems. A need is a necessary condition that must be met if a particular problem or set of problems is to be solved. It is possible that the needs statement will have to describe the essentiality for relaxing requirements and goal constraints governing the design.

Formulate Goals and Requirements

Mapping between the analysis model's function and use scenarios and the requirements and goal statements. When the design fails to adequately support activities and use scenarios, the requirements model allows the designer to easily identify the requirements constraints or goal statements that have been violated.

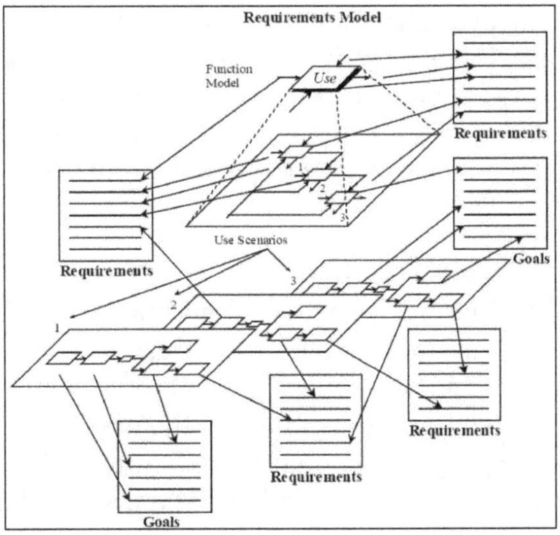

Once the needs for the design transition have been identified, the designer formulates:

- Requirements that the solution must satisfy.
- Goals that the solution should attempt to satisfy.

A requirement is a constraint on either the functional, behavioral, physical, or method of development aspects of a solution. A design goal is a stated aim that the design structure and specifications must support.

Formulate Solution Strategies

Once the requirements and goals have been established, the design team formulates alternative strategies for exploration in the next major transition in the design.

Design strategies can be considered as "meta-plans" for dealing with frequently occurring design situations. They can be viewed as methodizations or organizations of the primitive design activities identified above (i.e., partitioning, classification/specification, assembly, simulation, and re-partitioning). The three types of design strategies considered in the IDEF4 rationale component include:

- External-constraint-driven design: Design carried out under situations where the goals, intentions, and requirements are not characterized well, much less defined. These situations often result when the designer is brought into the product development process too early.
- Characteristic-driven design: Design in a closely controlled situation where strict accountability and proof of adequacy are rigidly enforced. These design situations often involve potentially life-threatening situations.
- Carry-over-driven design: Sometimes referred to as "routine" design.

Design as a cognitive endeavor shares many characteristics with other activities such as planning and diagnosis. But, design is distinguished by the context in which it is performed, the generic activities involved, the strategies employed, and the types of knowledge applied. A major distinguishing characteristic is the focus of the design process on the creation (refinement, analysis, etc.) of a specification of the end product.

IDEF8

IDEF8, or integrated definition for human-system interaction design, is a method for producing high-quality designs of interactions between users and the systems they operate. Systems are characterized as a collection of objects that perform functions to accomplish a particular goal. The system with which the user interacts can be any system, not necessarily a computer program. Human-system interactions are designed at three levels of specification within the IDEF8 method. The first level defines the philosophy of system operation and produces a set of models and textual descriptions of overall system processes. The second level of design specifies role-centered scenarios of system use. The third level of IDEF8 design is for human-system design detailing. At this level of design, IDEF8 provides a library of metaphors to help users and designers specify the desired behavior in terms of other objects whose behavior is more familiar. Metaphors provide a model of abstract concepts in terms of familiar, concrete objects and experiences.

IDEF9

IDEF9, or integrated definition for business constraint discovery, is designed to assist in the discovery and analysis of constraints in a business system. A primary motivation driving the development of IDEF9 was an acknowledgment that the collection of constraints that forge an enterprise system is generally poorly defined. The knowledge of what constraints exist and how those constraints interact is incomplete, disjoint, distributed, and often completely unknown. Just as living organisms do not need to be aware of the genetic or autonomous constraints that govern certain behaviors, organizations can (and most do) perform well without explicit knowledge of the glue that structures the system. In order to modify business in a predictable manner, however, the knowledge of these constraints is as critical as knowledge of genetics is to the genetic engineer.

IDEF14

IDEF14, or integrated definition for network design method, is a method that targets the modeling and design of computer and communication networks. It can be used to model existing ("as is") or envisioned ("to be") networks. It helps the network designer to investigate potential network designs and to document design rationale. The fundamental goals of the IDEF14 research project developed from a perceived need for good network designs that can be implemented quickly and accurately.

Diagrams in Systems Modeling Language

Package Diagram

A package diagram in the Unified Modeling Language depicts the dependencies between the packages that make up a model.

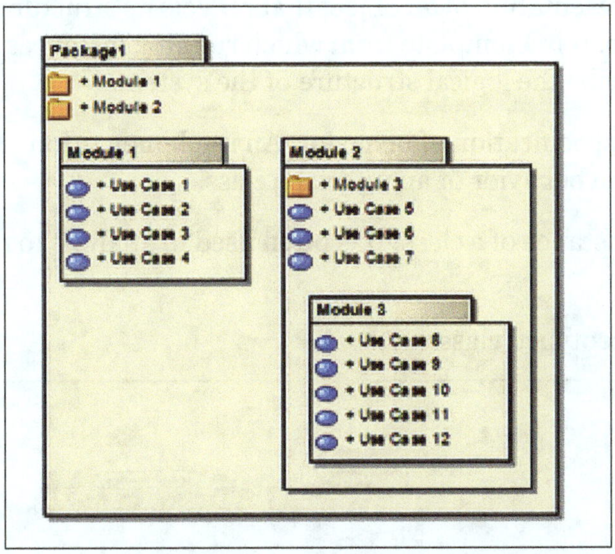

Packages containing nodes and artifacts.

In addition to the standard UML Dependency relationship, there are two special types of dependencies defined between packages:

- Package import,
- Package merge.

A package import is "a relationship between an importing namespace and a package, indicating that the importing namespace adds the names of the members of the package to its own namespace." By default, an unlabeled dependency between two packages is interpreted as a package import relationship. In this relationship, elements within the target package will be imported into the source package.

A package merge is "a directed relationship between two packages, that indicates that the contents of the two packages are to be combined. It is very similar to Generalisation in the sense that the source element conceptually adds the characteristics of the target element to its own characteristics resulting in an element that combines the characteristics of both" In this relationship, if an element exists within both the source package and the target package, then the source element's definition will be expanded to include the target element's definition.

Elements

- Package: A general purpose mechanism for organising model elements & diagrams into groups. It provides an encapsulated namespace within which all the names must be unique. It is used to group semantically related elements. It is a namespace as well as an element that can be contained in other packages' namespaces.
- Class: A representation of an object that reflects its structure and behavior within the system. It is a template from which running instances are created. Classes usually describe the logical structure of the system.
- Interface: A specification of behavior. An implementation class must be written to support the behavior of an interface class.
- Object: An instance of a class. It is often used in analysis to represent an artifact or other item.
- Table: A stereotyped class.

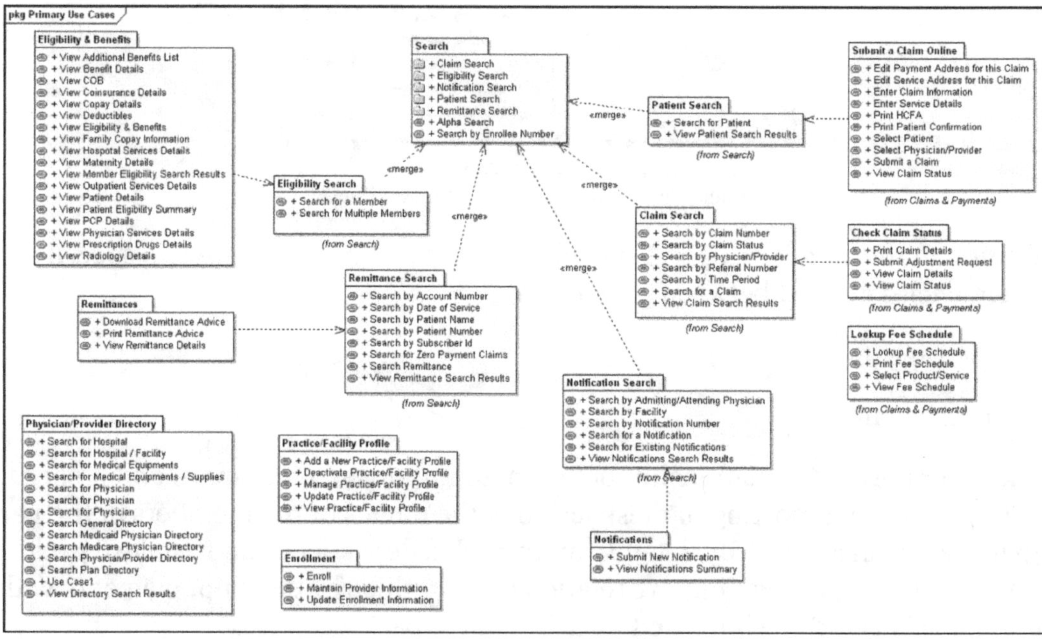

Packages containing use cases.

Usage

Package diagrams can use packages containing use cases to illustrate the functionality of a software system.

Package diagrams can use packages that represent the different layers of a software system to illustrate the layered architecture of a software system. The dependencies between these packages can be adorned with labels/stereotypes to indicate the communication mechanism between the layers.

- When To Use:
  - It is used in large scale systems to picture dependencies between major elements in the system.
  - Package diagrams represent a compile time grouping mechanism.

Use Case Diagram

A use case diagram at its simplest is a representation of a user's interaction with the system that shows the relationship between the user and the different use cases in which the user is involved. A use case diagram can identify the different types of users of a system and the different use cases and will often be accompanied by other types of diagrams as well. The use cases are represented by either circles or ellipses.

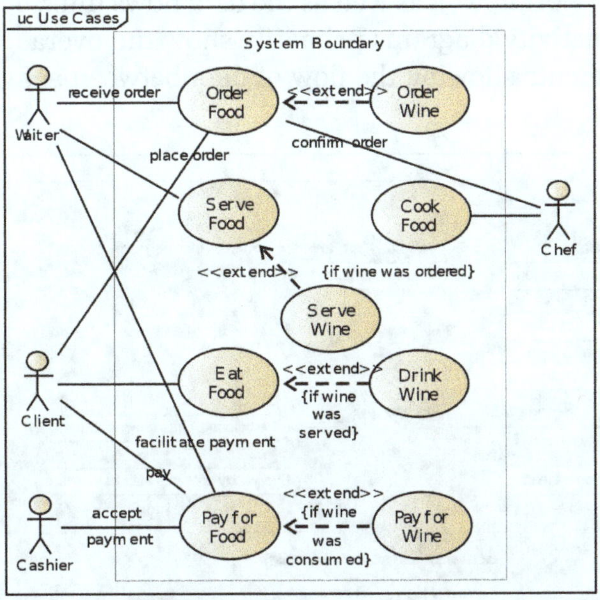

A UML use case diagram for the interaction of a client (the actor) within a restaurant (the system).

Application

While a use case itself might drill into a lot of detail about every possibility, a use-case diagram can help provide a higher-level view of the system. It has been said before that "Use case diagrams are the blueprints for your system". They provide the simplified and graphical representation of what the system must actually do.

Due to their simplistic nature, use case diagrams can be a good communication tool for stakeholders. The drawings attempt to mimic the real world and provide a view for the stakeholder to understand how the system is going to be designed. Siau and Lee conducted research to determine if there was a valid situation for use case diagrams at all or if they were unnecessary. What was found was that the use case diagrams conveyed the intent of the system in a more simplified manner to stakeholders and that they were "interpreted more completely than class diagrams".

The purpose of the use case diagrams is simply to provide the high level view of the system and convey the requirements in laypeople's terms for the stakeholders. Additional diagrams and documentation can be used to provide a complete functional and technical view of the system.

Activity Diagram

Activity diagrams are graphical representations of workflows of stepwise activities and actions with support for choice, iteration and concurrency. In the Unified Modeling Language, activity diagrams are intended to model both computational and organizational processes (i.e., workflows), as well as the data flows intersecting with the related activities. Although activity diagrams primarily show the overall flow of control, they can also include elements showing the flow of data between activities through one or more data stores.

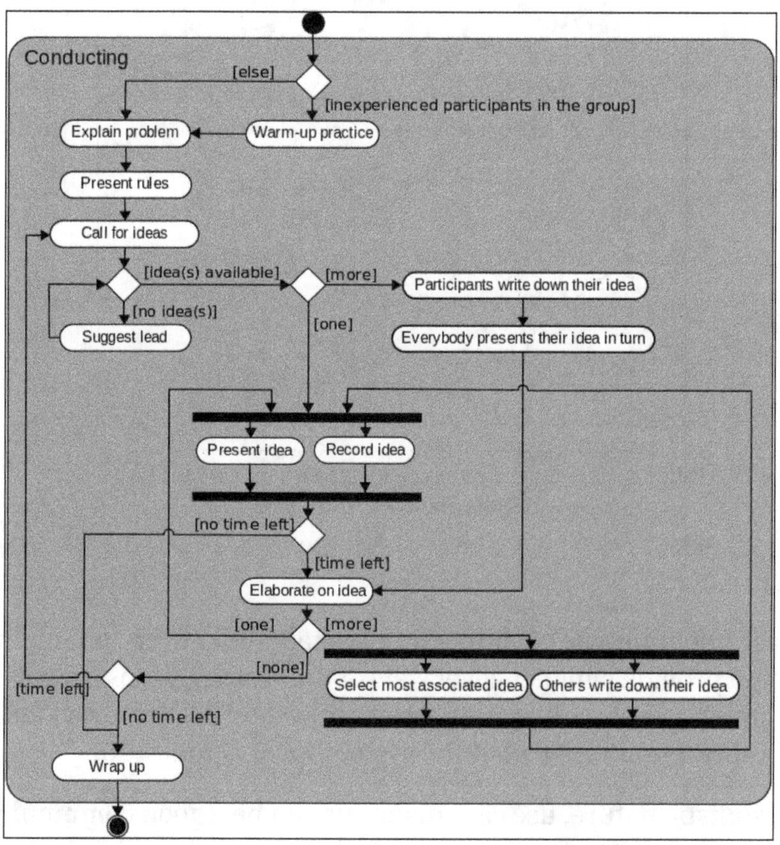

UML 1.x Activity diagram for a guided brainstorming process.

Construction

Activity diagrams are constructed from a limited number of shapes, connected with arrows. The most important shape types:

- Ellipses represent actions.
- Diamonds represent decisions.
- Bars represent the start (split) or end (join) of concurrent activities.
- A black circle represents the start (initial node) of the workflow.
- An encircled black circle represents the end (final node).

Arrows run from the start towards the end and represent the order in which activities happen.

Activity diagrams can be regarded as a form of a structured flowchart combined with a traditional data flow diagram. Typical flowchart techniques lack constructs for expressing concurrency. However, the join and split symbols in activity diagrams only resolve this for simple cases; the meaning of the model is not clear when they are arbitrarily combined with decisions or loops.

While in UML 1.x, activity diagrams were a specialized form of state diagrams, in UML 2.x, the activity diagrams were reformalized to be based on Petri net-like semantics, increasing the scope of situations that can be modeled using activity diagrams. These changes cause many UML 1.x activity diagrams to be interpreted differently in UML 2.x.

UML activity diagrams in version 2.x can be used in various domains, e.g. in design of embedded systems. It is possible to verify such a specification using model checking technique.

Sequence Diagram

A sequence diagram shows object interactions arranged in time sequence. It depicts the objects and classes involved in the scenario and the sequence of messages exchanged between the objects needed to carry out the functionality of the scenario. Sequence diagrams are typically associated with use case realizations in the Logical View of the system under development. Sequence diagrams are sometimes called event diagrams or event scenarios.

A sequence diagram shows, as parallel vertical lines (lifelines), different processes or objects that live simultaneously, and, as horizontal arrows, the messages exchanged between them, in the order in which they occur. This allows the specification of simple runtime scenarios in a graphical manner.

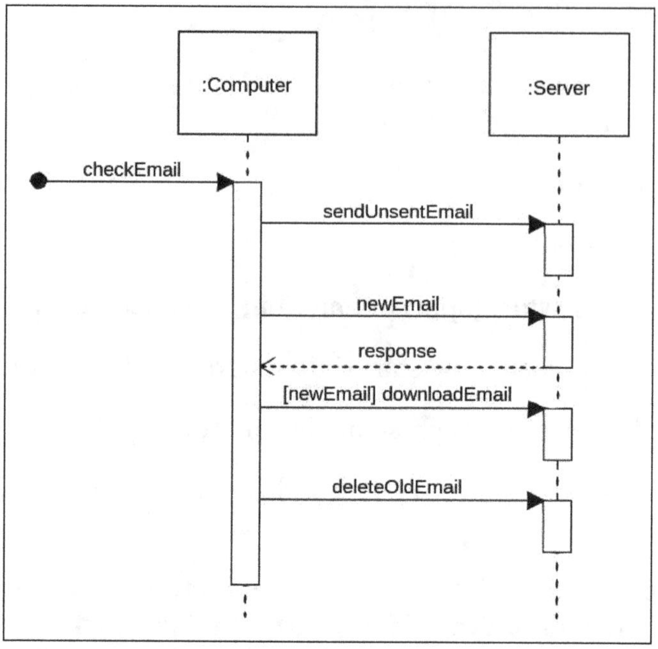

The Sequence diagram of UML.

Diagram Building Blocks

If the lifeline is that of an object, it demonstrates a role. Leaving the instance name blank can represent anonymous and unnamed instances.

Messages, written with horizontal arrows with the message name written above them, display interaction. Solid arrow heads represent synchronous calls, open arrow heads represent asynchronous messages, and dashed lines represent reply messages. If a caller sends a synchronous message, it must wait until the message is done, such as invoking a subroutine. If a caller sends an asynchronous message, it can continue processing and doesn't have to wait for a response. Asynchronous calls are present in multithreaded applications, event-driven applications and in message-oriented middleware. Activation boxes, or method-call boxes, are opaque rectangles drawn on top of lifelines to represent that processes are being performed in response to the message (ExecutionSpecifications in UML).

Objects calling methods on themselves use messages and add new activation boxes on top of any others to indicate a further level of processing. If an object is destroyed (removed from memory), an X is drawn on bottom of the lifeline, and the dashed line ceases to be drawn below it. It should be the result of a message, either from the object itself, or another.

A message sent from outside the diagram can be represented by a message originating from a filled-in circle (found message in UML) or from a border of the sequence diagram (gate in UML).

UML has introduced significant improvements to the capabilities of sequence diagrams. Most of these improvements are based on the idea of interaction fragments which represent smaller pieces of an enclosing interaction. Multiple interaction fragments are combined to create a variety of combined fragments, which are then used to model interactions that include parallelism, conditional branches, optional interactions.

State Diagram

A state diagram is a type of diagram used in computer science and related fields to describe the behavior of systems. State diagrams require that the system described is composed of a finite number of states; sometimes, this is indeed the case, while at other times this is a reasonable abstraction. Many forms of state diagrams exist, which differ slightly and have different semantics.

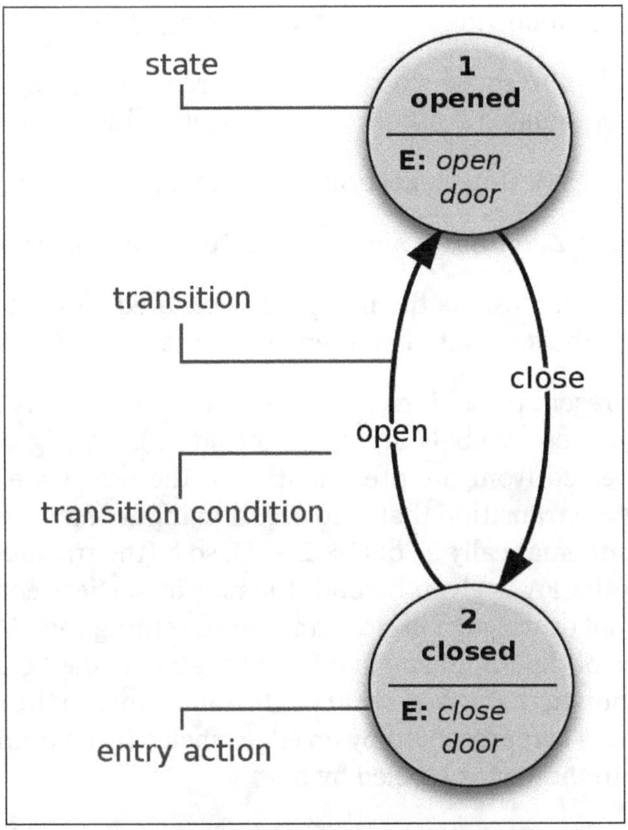

A state diagram for a door that can only be opened and closed.

State diagrams are used to give an abstract description of the behavior of a system. This behavior is analyzed and represented by a series of events that can occur in one or more possible states. Hereby "each diagram usually represents objects of a single class and track the different states of its objects through the system".

Directed Graph

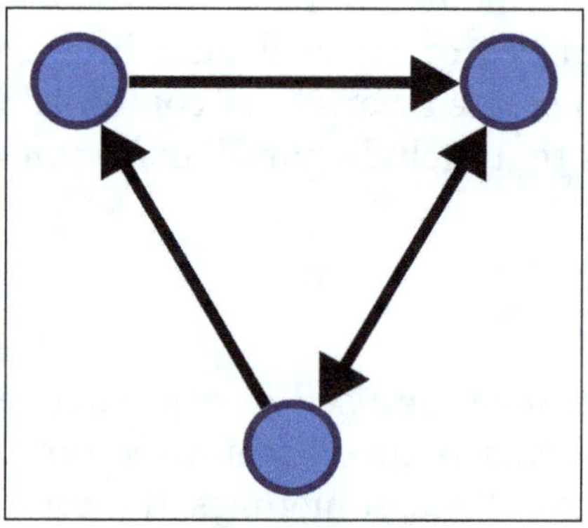

A directed graph.

A classic form of state diagram for a finite state machine or finite automaton (FA) is a directed graph with the following elements (Q,Σ,Z,δ,q_o,F):

- Vertices Q: A finite set of states, normally represented by circles and labeled with unique designator symbols or words written inside them.
- Input symbols Σ: A finite collection of input symbols or designators.
- Output symbols Z: A finite collection of output symbols or designators.

The output function ω represents the mapping of ordered pairs of input symbols and states onto output symbols, denoted mathematically as $\omega: \Sigma \times Q \rightarrow Z$.

- Edges δ: Represent transitions from one state to another as caused by the input (identified by their symbols drawn on the edges). An edge is usually drawn as an arrow directed from the present state to the next state. This mapping describes the state transition that is to occur on input of a particular symbol. This is written mathematically as $\delta: Q \times \Sigma \rightarrow Q$, so δ (the transition function) in the definition of the FA is given by both the pair of vertices connected by an edge and the symbol on an edge in a diagram representing this FA. Item $\delta(q,a)= p$ in the definition of the FA means that from the state named q under input symbol a, the transition to the state p occurs in this machine. In the diagram representing this FA, this is represented by an edge labeled by a pointing from the vertex labeled by q to the vertex labeled by p.
- Start state q_o: The start state $q_o \in Q$ is usually represented by an arrow with no origin pointing to the state. In older texts, the start state is not shown and must be inferred from the text.
- Accepting state(s) F: If used, for example for accepting automata, $F \in Q$ is the accepting state. It is usually drawn as a double circle. Sometimes the accept state(s) function as "Final" (halt, trapped) states.

For a deterministic finite automaton (DFA), nondeterministic finite automaton (NFA), generalized nondeterministic finite automaton (GNFA), or Moore machine, the input is denoted on each edge. For a Mealy machine, input and output are signified on each edge, separated with a slash "/": "1/0" denotes the state change upon encountering the symbol "1" causing the symbol "0" to be output. For a Moore machine the state's output is usually written inside the state's circle, also separated from the state's designator with a slash "/". There are also variants that combine these two notations.

For example, if a state has a number of outputs (e.g. "a= motor counter-clockwise=1, b= caution light inactive=0") the diagram should reflect this: e.g. "q5/1,0" designates state q5 with outputs a=1, b=0. This designator will be written inside the state's circle.

Example of DFA, NFA, GNFA or Moore Machine

S_1 and S_2 are states and S_1 is an accepting state or a final state. Each edge is labeled with the input. This example shows an acceptor for strings over {0,1} that contain an even number of zeros.

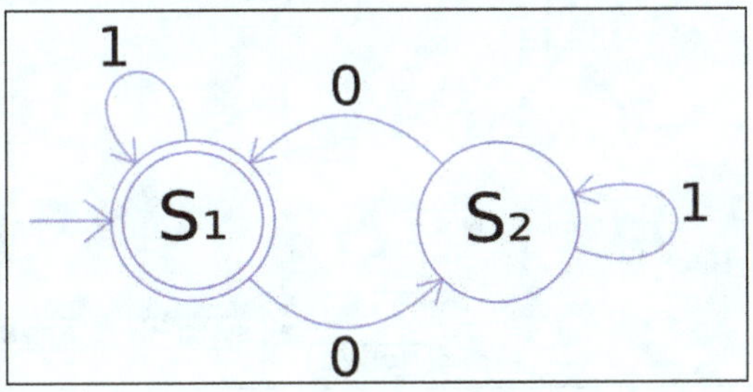

Example of Mealy Machine

S_0, S_1, and S_2 are states. Each edge is labeled with "j / k" where j is the input and k is the output.

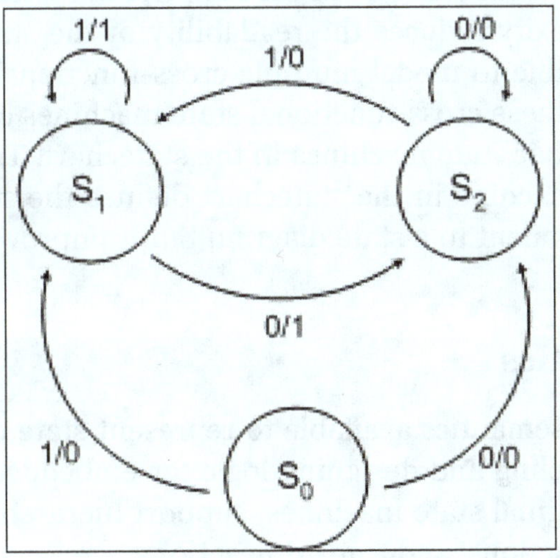

Harel Statechart

Harel statecharts, invented by computer scientist David Harel, are gaining widespread usage since a variant has become part of the Unified Modeling Language (UML). The diagram type allows the modeling of superstates, orthogonal regions, and activities as part of a state.

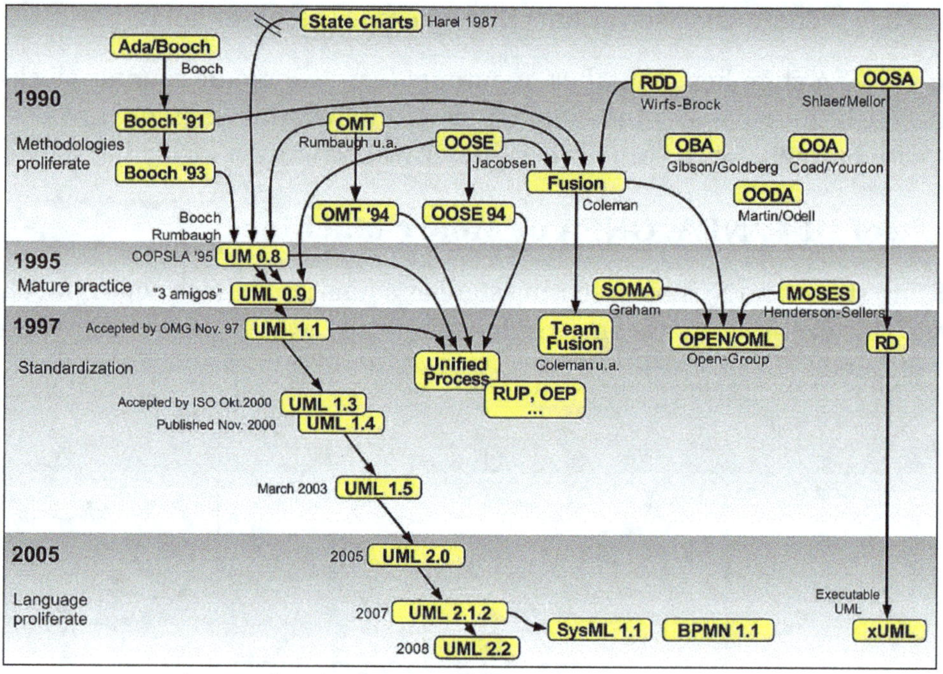

Diagram showing how Harel's Statecharts contributed to object-oriented methods and notation.

Classic state diagrams require the creation of distinct nodes for every valid combination of parameters that define the state. This can lead to a very large number of nodes and transitions between nodes for all but the simplest of systems (state and transition explosion). This complexity reduces the readability of the state diagram. With Harel statecharts it is possible to model multiple cross-functional state diagrams within the statechart. Each of these cross-functional state machines can transition internally without affecting the other state machines in the statechart. The current state of each cross-functional state machine in the statechart defines the state of the system. The Harel statechart is equivalent to a state diagram but it improves the readability of the resulting diagram.

Alternative Semantics

There are other sets of semantics available to represent state diagrams. For example, there are tools for modeling and designing logic for embedded controllers. These diagrams, like Harel's original state machines, support hierarchically nested states, orthogonal regions, state actions, and transition actions.

State Diagrams vs. Flowcharts

Newcomers to the state machine formalism often confuse state diagrams with flowcharts. The figure below shows a comparison of a state diagram with a flowchart. A state machine (panel (a)) performs actions in response to explicit events. In contrast, the flowchart (panel (b)) does not need explicit events but rather transitions from node to node in its graph automatically upon completion of activities.

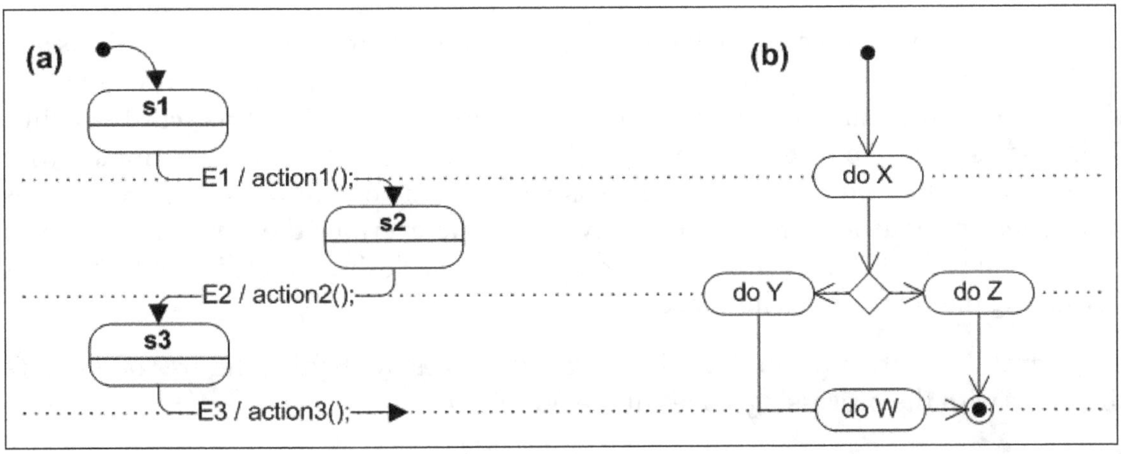

Nodes of flowcharts are edges in the induced graph of states. The reason is that each node in a flowchart represents a program command. A program command is an action to be executed. So it is not a state, but when applied to the program's state, it results in a transition to another state.

The source code listing represents a program graph. Executing the program graph (parsing and interpreting) results in a state graph. So each program graph induces a state graph. Conversion of the program graph to its associated state graph is called "unfolding" of the program graph. The program graph is a sequence of commands. If no variables exist, then the state consists only of the program counter, which keeps track of where in the program we are during execution (what is the next command to be applied).

In this case before executing a command the program counter is at some position (state before the command is executed). Executing the command moves the program counter to the next command. Since the program counter is the whole state, it follows that executing the command changed the state. So the command itself corresponds to a transition between the two states.

Now consider the full case, when variables exist and are affected by the program commands being executed. Then between different program counter locations, not only does the program counter change, but variables might also change values, due to the commands executed. Consequently, even if we revisit some program command (e.g. in a loop), this doesn't imply the program is in the same state.

In the previous case, the program would be in the same state, because the whole state is just the program counter, so if the program counterpoints to the same position (next command) it suffices to specify that we are in the same state. However, if the state includes variables, then if those change value, we can be at the same program location with different variable values, meaning in a different state in the program's state space. The term "unfolding" originates from this multiplication of locations when producing the state graph from the program graph.

A representative example is a do loop incrementing some counter until it overflows and becomes 0 again. Although the do loop executes the same increment command iteratively, so the program graph executes a cycle, in its state space is not a cycle, but a line. This results from the state being the program location (here cycling) combined with the counter value, which is strictly increasing (until the overflow), so different states are visited in sequence, until the overflow. After the overflow the counter becomes 0 again, so the initial state is revisited in the state space, closing a cycle in the state space (assuming the counter was initialized to 0).

The figure above attempts to show that reversal of roles by aligning the arcs of the state diagrams with the processing stages of the flowchart.

You can compare a flowchart to an assembly line in manufacturing because the flowchart describes the progression of some task from beginning to end (e.g., transforming source code input into object code output by a compiler). A state machine generally has no notion of such a progression. The door state machine is not in a more advanced stage when it is in the "closed" state, compared to being in the "opened" state; it simply reacts differently to the open/close events. A state in a state machine is an efficient way of specifying a particular behavior, rather than a stage of processing.

Other Extensions

An interesting extension is to allow arcs to flow from any number of states to any number of states. This only makes sense if the system is allowed to be in multiple states at once, which implies that an individual state only describes a condition or other partial aspect of the overall, global state. The resulting formalism is known as a Petri net. Another extension allows the integration of flowcharts within Harel statecharts. This extension supports the development of software that is both event driven and workflow driven.

Requirements Diagram

A SysML requirement diagram enables you to visualize any kind of requirements of your system, both functinal and non-functional. You can also visualize the inter-relationships between requirements. By using SysML requirement diagram with UeXceler, you have a complete set of system requirements that involve the business goal, the user stories that describe user problems/concerns and the requirements to address the problems.

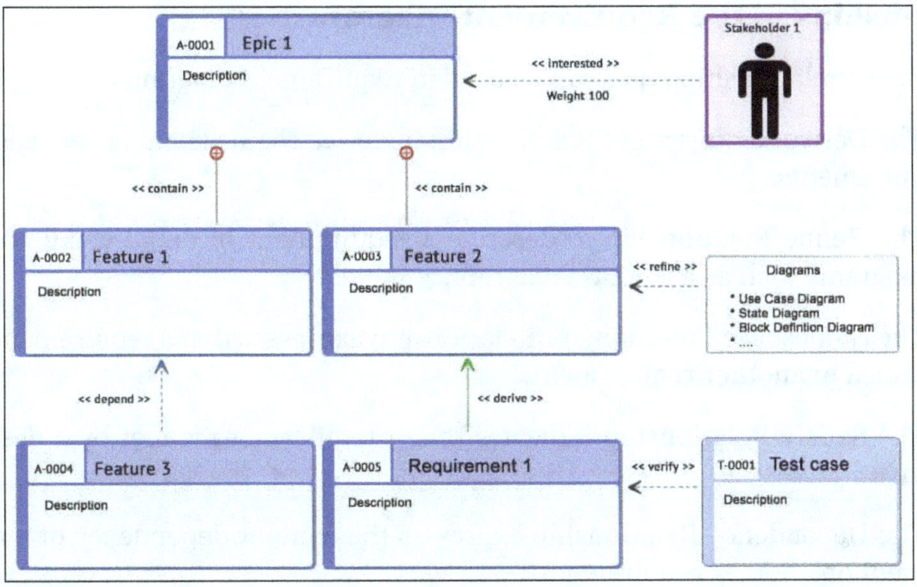

Ideal Number of Diagram Elements

Tests have shown that short sentences are easier to understand than long sentences. The best text comprehension is achieved with sentences of 9 words. So should diagrams also contain about 9 elements? If you develop systems, you might ask yourself such a question. But there is no formula or test to determine the ideal number of model elements per diagram, nor is there an optimal number of diagrams to describe systems. It is important that you use a requirement diagram to improve your understanding and make it easy for you to see what information is in the diagram or what is missing. If you develop the requirement diagram as part of a team or explain it to the team after it has been created, common understanding will be fostered and that is what modeling requirement diagrams is all about.

The Importance of Requirements

Developing a software or system without requirements is impossible. A system should do something and that something arises from the realisation of requirements. Stakeholders pursue goals and they expect the developed solution to help them achieve their goals. However, goals are too rough to derive specific characteristics of a system from them – requirements help here. Requirements concretise goals. Even when using abstractions such as Epics and Features, the message is the same: Requirements are essential for the development of a solution, especially since implementation activities, risk assessment or architectural design are based on them. Two points are important: Requirements for a new system must be complete, consistent and precise. Ideally, requirements can be determined using a defined process. And dependencies and relationships must be as clear as possible before implementation – requirement diagrams help here.

Relationships in the Requirement Diagram

The following relationship types can be used in requirement diagrams:

- The Derive Requirement Relationship to derive requirements from existing requirements.
- The Refine Relationship to describe a requirement in detail using additional diagrams such as a use case diagram.
- The Namespace Containment Relationship expresses that a requirement is contained in another requirement.
- The Satisfy Relationship indicates that a requirement is met by a design element.
- The Dependency Relationship expresses the content dependency of a requirement on another requirement.
- The Verify Relationship to verify a requirement by a test case.
- The Copy Relationship identifies the relationship between a requirement copy and its original.
- The Trace Relationship can be used to track relationships, for example, to link functional to non-functional requirements.

Some tool vendors offer the use of additional relationships between elements in the requirement diagram:

- The Interest Relationship s used when, for example, a stakeholder or persona is interested in a requirement.
- The Needs Relationship is used when a stakeholder or persona needs a solution to a requirement.
- The Note Relationship helps to display additional explanations or notes.

Advantages of Requirement Diagrams

Working with requirement diagrams offers several advantages:

- They are easy to create and understand. Communication about concrete information is made easier and correlations are identified more easily.
- Missing aspects can be easily identified to help identify additional requirements.
- The different types of relationships facilitate traceability and promote understanding.

- They provide for order and thus for a higher quality when working with requirements.
- Compared to the pure textual description of requirements, they facilitate detailed work on concrete aspects that are important for development.

Example:

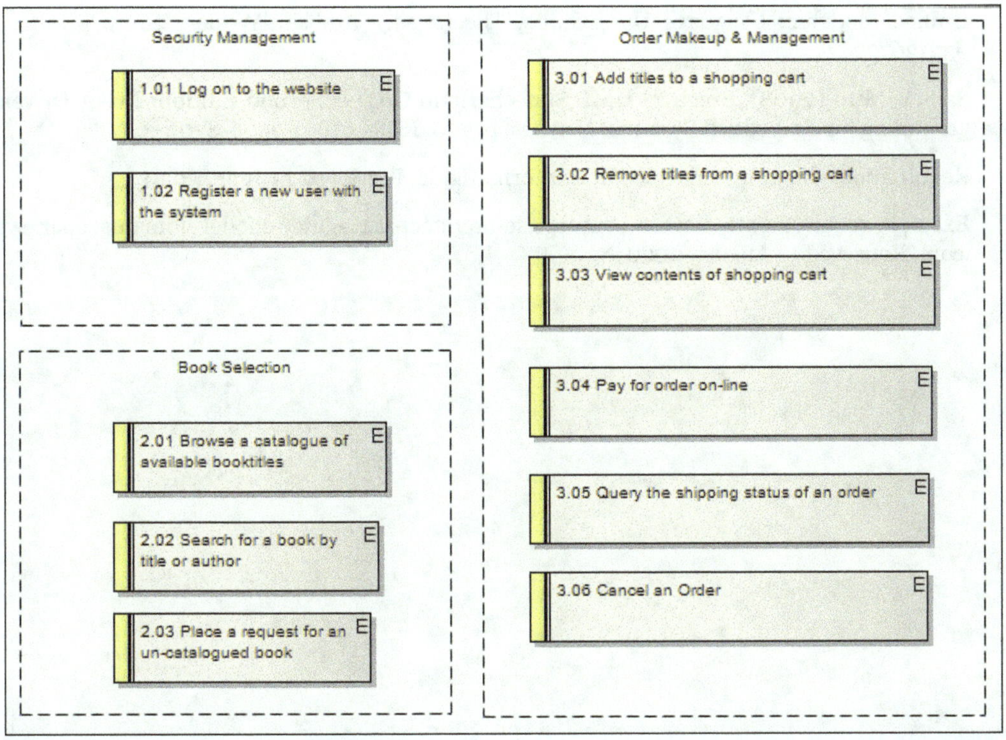

Example:

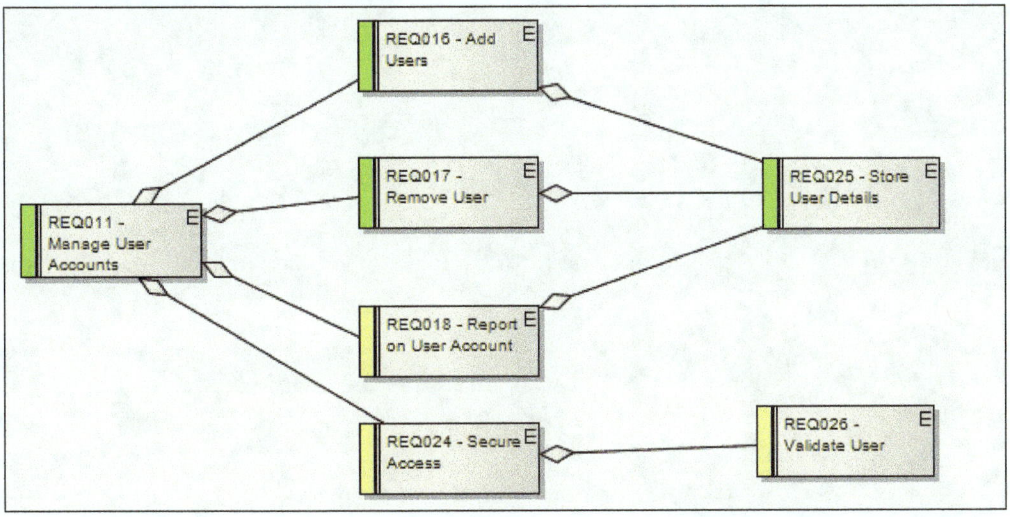

Example of Requirements Diagram: These examples illustrate possible structures of a Requirements diagram. Use Cases and Components in the system can be linked back to the Requirement elements to define how a particular system requirement is met.

References

- What-is-sysml-2: dthomas-software.co.uk, Retrieved 28 May, 2020
- Unified Modeling Language User Guide, The (2 ed.). Addison-Wesley. 2005. p. 496. ISBN 0321267974
- Samek, Miro (2008). Practical UML Statecharts in C/C++, Second Edition: Event-Driven Programming for Embedded Systems. Newnes. p. 728. ISBN 978-0-7506-8706-5
- Requirement-diagram, smartpedia: t2informatik.de, Retrieved 13 June, 2020
- Example-requirements-diagram, enterprise-architect-user-guide-model-domains: sparxsystems .com, Retrieved 17 August, 2020

5

Control Engineering

Control engineering is a discipline of engineering that is concerned with the application of automatic control theory to design systems in control environments. It includes control theory, stochastic control, open and closed loop system, feedback systems, etc. This chapter discusses the discipline of control engineering in detail.

Control system engineering is the branch of engineering which deals with the principles of control theory, to design a system which gives yields the desired behavior in a controlled manner. Hence, although control engineering is often taught within electrical engineering at university, it is an interdisciplinary topic.

Control system engineers analyze, design, and optimize complex systems which consist of highly integrated coordination of mechanical, electrical, chemical, metallurgical, electronic or pneumatic elements. Thus control engineering deals with a diverse range of dynamic systems which include human and technological interfacing. These systems are broadly referred to as control systems.

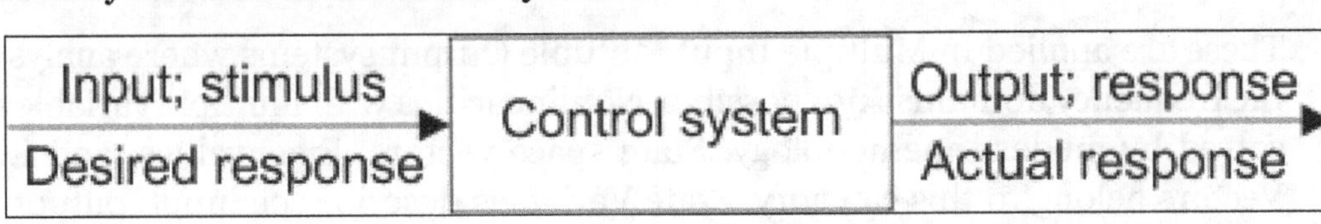

Control system engineering focuses on the analysis and design of systems to improve the speed of response, accuracy, and stability of the system. The two methods of control system include classical methods and modern methods. The mathematical model of the system is set up as the first step followed by analysis, designing and testing. Necessary conditions for the stability are checked and finally, optimization follows.

In the classical method, mathematical modeling is usually done in the time domain, frequency domain or complex domain. The step response of a system is mathematically modeled in time domain differential analysis to find its settling time, % overshoot, etc. Laplace transforms are most commonly used in the frequency domain to find the open loop gain, phase margin, bandwidth etc. of the system. The concept of the transfer function, Nyquist stability criteria, sampling of data, Nyquist plot, poles and zeros, Bode plots, system delays all come under the umbrella of classical control engineering stream.

Modern control engineering deals with Multiple Input Multiple Output (MIMO) systems, State space approach, Eigenvalues, and vectors, etc. Instead of transforming complex ordinary differential equations, modern approach converts higher order equations to first order differential equations and solved by vector method.

Automatic control systems are most commonly used as it does not involve manual control. The controlled variable is measured and compared with a specified value to obtain the desired result. As a result of automated systems for control purposes, the cost of energy or power, as well as the cost of the process, will be reduced increasing its quality and productivity.

Types of Control Engineering

Control engineering has its own categorization depending on the different methodologies used. The main types of control engineering include:

- Classical Control Engineering: The systems are usually represented by using ordinary differential equations. In classical control engineering, these equations are transformed and analyzed in a transformed domain. Laplace transform, Fourier transform and z transform are examples. This method is commonly used in Single Input Single Output systems (SISO).
- Modern Control Engineering: In modern control engineering, higher order differential equations are converted to first order differential equations. These equations are solved very similar to vector method. By doing so, many complications dealt in solving higher order differential equations are solved.
- These are applied in Multiple Input Multiple Output systems where analysis in the frequency domain is not possible. Nonlinearities with multiple variables are solved by modern methodology. State space vectors, Eigenvalues, and Eigen Vectors belong to this category. State Variables describe the input, output and system variables.
- Robust Control Engineering: In robust control methodology, the changes in the performance of the system with the change in parameters are measured for optimization. This aids in widening the stability and performance, also in finding alternate solutions. Hence in robust control, the environment, internal inaccuracies, noises, and disturbances are considered to reduce the fault in the system.
- Optimal Control Engineering: In optimal control engineering, the problem is formulated as a mathematical model of the process, physical constraints and performance constraints, to minimize the cost function. Thus optimal control engineering is the most feasible solution for designing a system with minimum cost.

- Adaptive Control Engineering: In adaptive control engineering, the controllers employed are adaptive controllers in which parameters are made adaptive by some mechanism. The block diagram given below shows an adaptive control system.

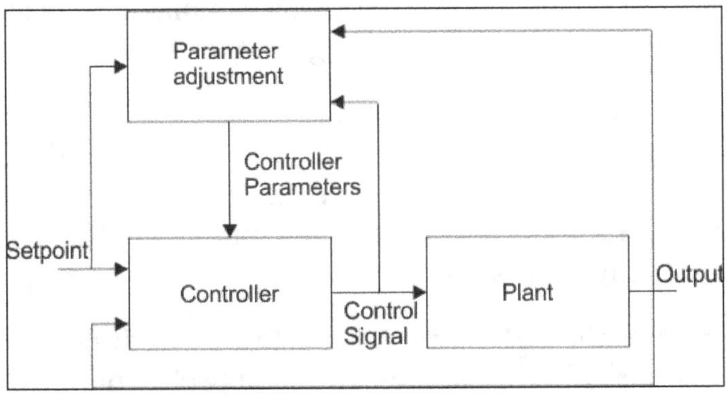

In this kind of controllers, an additional loop for parameter adjustment is present in addition to the normal feedback of process.

- Nonlinear Control Engineering: Nonlinear control engineering focuses on the nonlinearities which cannot be represented by using linear ordinary differential equations (i.e. they are not linear control systems). This system will exhibit multiple isolated equilibrium points, limit cycles, bifurcations with finite escape time. The main limitation is that it requires laborious mathematical analysis. In this analysis, the system is divided into the linear part and the nonlinear part.

- Game Theory: In game theory, each system will have to reduce its cost function against the disturbances/noises. Hence it is a study of conflict and cooperation. The disturbances will try to maximize the cost function. This theory is related to robust and optimal control engineering.

Control Theory

Control theory in control systems engineering is a subfield of mathematics that deals with the control of continuously operating dynamical systems in engineered processes and machines. The objective is to develop a control model for controlling such systems using a control action in an optimum manner without *delay or overshoot* and ensuring control stability.

To do this, a controller with the requisite corrective behaviour is required. This controller monitors the controlled process variable (PV), and compares it with the reference or set point (SP). The difference between actual and desired value of the process variable, called the *error* signal, or SP-PV error, is applied as feedback to generate a control action to bring the controlled process variable to the same value as the set point. Other aspects which are also studied are controllability and observability. This is the basis for the advanced type of automation that revolutionized manufacturing, aircraft, communications and other industries. This is feedback control, which is usually continuous and involves taking measurements using a sensor and making calculated adjustments to keep the measured variable within a set range by means of a "final control element", such as a control valve.

Extensive use is usually made of a diagrammatic style known as the block diagram. In it the transfer function, also known as the system function or network function, is a mathematical model of the relation between the input and output based on the differential equations describing the system.

Control theory dates from the 19th century, when the theoretical basis for the operation of governors was first described by James Clerk Maxwell. Control theory was further advanced by Edward Routh in 1874, Charles Sturm and in 1895, Adolf Hurwitz, who all contributed to the establishment of control stability criteria; and from 1922 onwards, the development of PID control theory by Nicolas Minorsky. Although a major application of control theory is in control systems engineering, which deals with the design of process control systems for industry, other applications range far beyond this. As the general theory of feedback systems, control theory is useful wherever feedback occurs.

PID Feedback Control

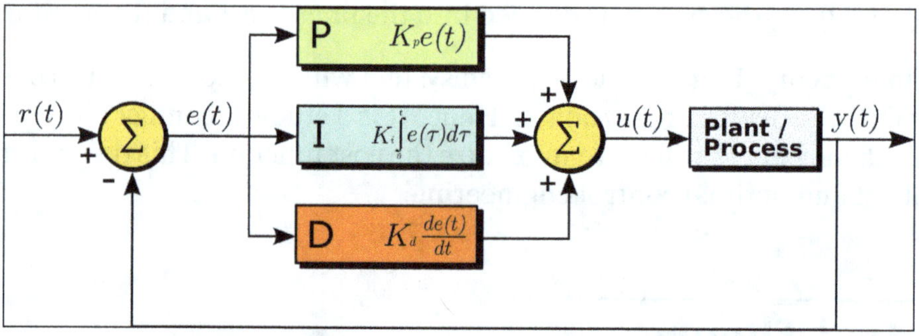

A block diagram of a PID controller in a feedback loop, r(*t*) is the desired process value or "set point", and y(*t*) is the measured process value.

A proportional–integral–derivative controller (PID controller) is a control loop feedback mechanism control technique widely used in control systems.

A PID controller continuously calculates an error value $e(t)$ as the difference between a desired setpoint and a measured process variable and applies a correction based on proportional, integral, and derivative terms. PID is an initialism for Proportional-Integral-Derivative, referring to the three terms operating on the error signal to produce a control signal.

The theoretical understanding and application dates from the 1920s, and they are implemented in nearly all analogue control systems; originally in mechanical controllers, and then using discrete electronics and latterly in industrial process computers. The PID controller is probably the most-used feedback control design.

If *u(t)* is the control signal sent to the system, *y(t)* is the measured output and *r(t)* is the desired output, and $e(t) = r(t) - y(t)$ is the tracking error, a PID controller has the general form,

$$(t) = K_P e(t) + K_I \int e(\tau) d\tau + K_D \frac{de(t)}{dt}.$$

The desired closed loop dynamics is obtained by adjusting the three parameters K_P, K_I and K_D, often iteratively by "tuning" and without specific knowledge of a plant model. Stability can often be ensured using only the proportional term. The integral term permits the rejection of a step disturbance (often a striking specification in process control). The derivative term is used to provide damping or shaping of the response. PID controllers are the most well-established class of control systems: however, they cannot be used in several more complicated cases, especially if MIMO systems are considered.

Applying Laplace transformation results in the transformed PID controller equation,

$$u(s) = K_P e(s) + K_I \frac{1}{s} e(s) + K_D s e(s)$$

$$u(s) = (K_P + K_I \frac{1}{s} + K_D s) e(s)$$

with the PID controller transfer function,

$$C(s) = \left(K_P + K_I \frac{1}{s} + K_D s \right).$$

As an example of tuning a PID controller in the closed-loop system $H(s)$, consider a 1st order plant given by,

$$P(s) = \frac{A}{1 + sT_P}$$

where A and T_P are some constants. The plant output is fed back through,

$$F(s) = \frac{1}{1 + sT_F}$$

where T_F is also a constant. Now if we set,

$K_P = K\left(1 + \frac{T_D}{T_I}\right)$, $K_D = KT_D$, and $K_I = \frac{K}{T_I}$, we can express the PID controller transfer function in series form as,

$$C(s) = K\left(1 + \frac{1}{sT_I}\right)(1 + sT_D)$$

Plugging $P(s)$, $F(s)$ and $C(s)$ into the closed-loop transfer function $H(s)$, we find that by setting,

$$K = \frac{1}{A}, T_I = T_F, T_D = T_P$$

$H(s) = 1$. With this tuning in this example, the system output follows the reference input exactly.

However, in practice, a pure differentiator is neither physically realizable nor desirable due to amplification of noise and resonant modes in the system. Therefore, a phase-lead compensator type approach or a differentiator with low-pass roll-off are used instead.

Linear and Nonlinear Control Theory

The field of control theory can be divided into two branches:

- Linear control theory: This applies to systems made of devices which obey the superposition principle, which means roughly that the output is proportional to the input. They are governed by linear differential equations. A major subclass is systems which in addition have parameters which do not change with time, called linear time invariant (LTI) systems. These systems are amenable to powerful frequency domain mathematical techniques of great generality, such as the Laplace transform, Fourier transform, Z transform, Bode plot, root locus, and Nyquist stability criterion. These lead to a description of the system using terms like bandwidth, frequency response, eigenvalues, gain, resonant frequencies, zeros and poles, which give solutions for system response and design techniques for most systems of interest.
- Nonlinear control theory: This covers a wider class of systems that do not obey the superposition principle, and applies to more real-world systems because all real control systems are nonlinear. These systems are often governed by nonlinear differential equations. The few mathematical techniques which have been developed to handle them are more difficult and much less general, often applying only to narrow categories of systems. These include limit cycle theory, Poincaré maps, Lyapunov stability theorem, and describing functions. Nonlinear systems are often analyzed using numerical methods on computers, for example by simulating their operation using a simulation language. If only solutions near a stable point are of interest, nonlinear systems can often be linearized by approximating them by a linear system using perturbation theory, and linear techniques can be used.

Analysis Techniques-frequency Domain and Time Domain

Mathematical techniques for analyzing and designing control systems fall into two different categories:

- Frequency domain: In this type the values of the state variables, the mathematical variables representing the system's input, output and feedback are represented as functions of frequency. The input signal and the system's transfer function are converted from time functions to functions of frequency by a transform such as the Fourier transform, Laplace transform, or Z transform. The advantage of this technique is that it results in a simplification of the mathematics; the differential equations that represent the system are replaced by algebraic equations in the frequency domain which is much simpler to solve. However, frequency domain techniques can only be used with linear systems.

- Time-domain state space representation: In this type the values of the state variables are represented as functions of time. With this model, the system being analyzed is represented by one or more differential equations. Since frequency domain techniques are limited to linear systems, time domain is widely used to analyze real-world nonlinear systems. Although these are more difficult to solve, modern computer simulation techniques such as simulation languages have made their analysis routine.

In contrast to the frequency domain analysis of the classical control theory, modern control theory utilizes the time-domain state space representation, a mathematical model of a physical system as a set of input, output and state variables related by first-order differential equations. To abstract from the number of inputs, outputs, and states, the variables are expressed as vectors and the differential and algebraic equations are written in matrix form (the latter only being possible when the dynamical system is linear). The state space representation (also known as the "time-domain approach") provides a convenient and compact way to model and analyze systems with multiple inputs and outputs. With inputs and outputs, we would otherwise have to write down Laplace transforms to encode all the information about a system. Unlike the frequency domain approach, the use of the state-space representation is not limited to systems with linear components and zero initial conditions. "State space" refers to the space whose axes are the state variables. The state of the system can be represented as a point within that space.

System Interfacing - SISO and MIMO

Control systems can be divided into different categories depending on the number of inputs and outputs:

- Single-input single-output (SISO): This is the simplest and most common type, in which one output is controlled by one control signal. Examples are the cruise control example above, or an audio system, in which the control input is the input audio signal and the output is the sound waves from the speaker.
- Multiple-input multiple-output (MIMO): These are found in more complicated systems. For example, modern large telescopes such as the Keck and MMT have mirrors composed of many separate segments each controlled by an actuator. The shape of the entire mirror is constantly adjusted by a MIMO active optics

control system using input from multiple sensors at the focal plane, to compensate for changes in the mirror shape due to thermal expansion, contraction, stresses as it is rotated and distortion of the wavefront due to turbulence in the atmosphere. Complicated systems such as nuclear reactors and human cells are simulated by a computer as large MIMO control systems.

Topics in Control Theory

Stability

The stability of a general dynamical system with no input can be described with Lyapunov stability criteria.

- A linear system is called bounded-input bounded-output (BIBO) stable if its output will stay bounded for any bounded input.
- Stability for nonlinear systems that take an input is input-to-state stability (ISS), which combines Lyapunov stability and a notion similar to BIBO stability.

For simplicity, the following descriptions focus on continuous-time and discrete-time linear systems.

Mathematically, this means that for a causal linear system to be stable all of the poles of its transfer function must have negative-real values, i.e. the real part of each pole must be less than zero. Practically speaking, stability requires that the transfer function complex poles reside:

- In the open left half of the complex plane for continuous time, when the Laplace transform is used to obtain the transfer function.
- Inside the unit circle for discrete time, when the Z-transform is used.

The difference between the two cases is simply due to the traditional method of plotting continuous time versus discrete time transfer functions. The continuous Laplace transform is in Cartesian coordinates where the x axis is the real axis and the discrete Z-transform is in circular coordinates where the ρ axis is the real axis.

When the appropriate conditions above are satisfied a system is said to be asymptotically stable; the variables of an asymptotically stable control system always decrease from their initial value and do not show permanent oscillations. Permanent oscillations occur when a pole has a real part exactly equal to zero (in the continuous time case) or a modulus equal to one (in the discrete time case). If a simply stable system response neither decays nor grows over time, and has no oscillations, it is marginally stable; in this case the system transfer function has non-repeated poles at the complex plane origin (i.e. their real and complex component is zero in the continuous time case). Oscillations are present when poles with real part equal to zero have an imaginary part not equal to zero.

If a system in question has an impulse response of,

$$x[n] = 0.5^n u[n]$$

then the Z-transform, is given by,

$$X(z) = \frac{1}{1 - 0.5z^{-1}}$$

which has a pole in $z = 0.5$ (zero imaginary part). This system is BIBO (asymptotically) stable since the pole is *inside* the unit circle.

However, if the impulse response was,

$$x[n] = 1.5^n u[n]$$

then the Z-transform is,

$$X(z) = \frac{1}{1 - 1.5z^{-1}}$$

which has a pole at $z = 1.5$ and is not BIBO stable since the pole has a modulus strictly greater than one.

Numerous tools exist for the analysis of the poles of a system. These include graphical systems like the root locus, Bode plots or the Nyquist plots.

Mechanical changes can make equipment (and control systems) more stable. Sailors add ballast to improve the stability of ships. Cruise ships use antiroll fins that extend transversely from the side of the ship for perhaps 30 feet (10 m) and are continuously rotated about their axes to develop forces that oppose the roll.

Controllability and Observability

Controllability and observability are main issues in the analysis of a system before deciding the best control strategy to be applied, or whether it is even possible to control or stabilize the system. Controllability is related to the possibility of forcing the system into a particular state by using an appropriate control signal. If a state is not controllable, then no signal will ever be able to control the state. If a state is not controllable, but its dynamics are stable, then the state is termed stabilizable. Observability instead is related to the possibility of observing, through output measurements, the state of a system. If a state is not observable, the controller will never be able to determine the behavior of an unobservable state and hence cannot use it to stabilize the system. However, similar to the stabilizability condition above, if a state cannot be observed it might still be detectable.

From a geometrical point of view, looking at the states of each variable of the system to be controlled, every "bad" state of these variables must be controllable and observable to ensure a good behavior in the closed-loop system. That is, if one of the eigenvalues of the system is not both controllable and observable, this part of the dynamics will remain untouched in the closed-loop system. If such an eigenvalue is not stable, the dynamics of this eigenvalue will be present in the closed-loop system which therefore will be unstable. Unobservable poles are not present in the transfer function realization of a state-space representation, which is why sometimes the latter is preferred in dynamical systems analysis.

Solutions to problems of an uncontrollable or unobservable system include adding actuators and sensors.

Control Specification

Several different control strategies have been devised in the past years. These vary from extremely general ones (PID controller), to others devoted to very particular classes of systems (especially robotics or aircraft cruise control).

A control problem can have several specifications. Stability, of course, is always present. The controller must ensure that the closed-loop system is stable, regardless of the open-loop stability. A poor choice of controller can even worsen the stability of the open-loop system, which must normally be avoided. Sometimes it would be desired to obtain particular dynamics in the closed loop: i.e. that the poles have $Re[\lambda] < -\bar{\lambda}$, where $\bar{\lambda}$ is a fixed value strictly greater than zero, instead of simply asking that $Re[\lambda] < 0$.

Another typical specification is the rejection of a step disturbance; including an integrator in the open-loop chain (i.e. directly before the system under control) easily achieves this. Other classes of disturbances need different types of sub-systems to be included.

Other "classical" control theory specifications regard the time-response of the closed-loop system. These include the rise time (the time needed by the control system to reach the desired value after a perturbation), peak overshoot (the highest value reached by the response before reaching the desired value) and others (settling time, quarter-decay). Frequency domain specifications are usually related to robustness. Modern performance assessments use some variation of integrated tracking error (IAE,ISA,CQI).

Model Identification and Robustness

A control system must always have some robustness property. A robust controller is such that its properties do not change much if applied to a system slightly different from the mathematical one used for its synthesis. This requirement is important, as no real physical system truly behaves like the series of differential equations used to represent it mathematically. Typically a simpler mathematical model is chosen in order to simplify calculations, otherwise, the true system dynamics can be so complicated that a complete model is impossible.

System Identification

The process of determining the equations that govern the model's dynamics is called system identification. This can be done off-line: for example, executing a series of measures from which to calculate an approximated mathematical model, typically its transfer function or matrix. Such identification from the output, however, cannot take account of unobservable dynamics. Sometimes the model is built directly starting from known physical equations, for example, in the case of a mass-spring-damper system we know that $m\ddot{x}(t) = -Kx(t) - B\dot{x}(t)$. Even assuming that a "complete" model is used in designing the controller, all the parameters included in these equations (called "nominal parameters") are never known with absolute precision; the control system will have to behave correctly even when connected to a physical system with true parameter values away from nominal.

Some advanced control techniques include an "on-line" identification process. The parameters of the model are calculated ("identified") while the controller itself is running. In this way, if a drastic variation of the parameters ensues, for example, if the robot's arm releases a weight, the controller will adjust itself consequently in order to ensure the correct performance.

Analysis

Analysis of the robustness of a SISO (single input single output) control system can be performed in the frequency domain, considering the system's transfer function and using Nyquist and Bode diagrams. Topics include gain and phase margin and amplitude margin. For MIMO (multi-input multi output) and, in general, more complicated control systems, one must consider the theoretical results devised for each control technique. i.e., if particular robustness qualities are needed, the engineer must shift his attention to a control technique by including them in its properties.

Constraints

A particular robustness issue is the requirement for a control system to perform properly in the presence of input and state constraints. In the physical world every signal is limited. It could happen that a controller will send control signals that cannot be followed by the physical system, for example, trying to rotate a valve at excessive speed. This can produce undesired behavior of the closed-loop system, or even damage or break actuators or other subsystems. Specific control techniques are available to solve the problem: model predictive control, and anti-wind up systems. The latter consists of an additional control block that ensures that the control signal never exceeds a given threshold.

Main Control Strategies

Every control system must guarantee first the stability of the closed-loop behavior. For linear systems, this can be obtained by directly placing the poles. Non-linear control systems use specific theories (normally based on Aleksandr Lyapunov's Theory) to ensure stability without regard to the inner dynamics of the system. The possibility to fulfill different specifications varies from the model considered and the control strategy chosen.

List of the main control techniques:

- Adaptive control uses on-line identification of the process parameters, or modification of controller gains, thereby obtaining strong robustness properties. Adaptive controls were applied for the first time in the aerospace industry in the 1950s, and have found particular success in that field.
- A hierarchical control system is a type of control system in which a set of devices and governing software is arranged in a hierarchical tree. When the links in the tree are implemented by a computer network, then that hierarchical control system is also a form of networked control system.
- Intelligent control uses various AI computing approaches like artificial neural networks, Bayesian probability, fuzzy logic, machine learning, evolutionary computation and genetic algorithms to control a dynamic system.
- Optimal control is a particular control technique in which the control signal optimizes a certain "cost index": for example, in the case of a satellite, the jet thrusts needed to bring it to desired trajectory that consume the least amount of fuel. Two optimal control design methods have been widely used in industrial applications, as it has been shown they can guarantee closed-loop stability. These are Model Predictive Control (MPC) and linear-quadratic-Gaussian control (LQG). The first can more explicitly take into account constraints on the signals in the system, which is an important feature in many industrial processes. However, the "optimal control" structure in MPC is only a means to achieve such a result, as it does not optimize a true performance index of the closed-loop control system. Together with PID controllers, MPC systems are the most widely used control technique in process control.
- Robust control deals explicitly with uncertainty in its approach to controller design. Controllers designed using robust control methods tend to be able to cope with small differences between the true system and the nominal model used for design. The early methods of Bode and others were fairly robust; the state-space methods invented in the 1960s and 1970s were sometimes found to lack robustness. Examples of modern robust control techniques include H-infinity loop-shaping developed by Duncan McFarlane and Keith Glover, Sliding mode control (SMC) developed by Vadim Utkin, and safe protocols designed for control of large heterogeneous populations of electric loads in Smart Power Grid applications. Robust methods aim to achieve robust performance and/or stability in the presence of small modeling errors.

- Stochastic control deals with control design with uncertainty in the model. In typical stochastic control problems, it is assumed that there exist random noise and disturbances in the model and the controller, and the control design must take into account these random deviations.
- Energy-shaping control view the plant and the controller as energy-transformation devices. The control strategy is formulated in terms of interconnection (in a power-preserving manner) in order to achieve a desired behavior.
- Self-organized criticality control may be defined as attempts to interfere in the processes by which the self-organized system dissipates energy.

Classical Control Theory

Classical control theory is a branch of control theory that deals with the behavior of dynamical systems with inputs, and how their behavior is modified by feedback, using the Laplace transform as a basic tool to model such systems.

The usual objective of control theory is to control a system, often called the *plant*, so its output follows a desired control signal, called the *reference*, which may be a fixed or changing value. To do this a *controller* is designed, which monitors the output and compares it with the reference. The difference between actual and desired output, called the *error* signal, is applied as feedback to the input of the system, to bring the actual output closer to the reference.

Classical control theory deals with linear time-invariant single-input single-output systems. The Laplace transform of the input and output signal of such systems can be calculated. The transfer function relates the Laplace transform of the input and the output.

Feedback

To overcome the limitations of the open-loop controller, classical control theory introduces feedback. A closed-loop controller uses feedback to control states or outputs of a dynamical system. Its name comes from the information path in the system: process inputs (e.g., voltage applied to an electric motor) have an effect on the process outputs (e.g., speed or torque of the motor), which is measured with sensors and processed by the controller; the result (the control signal) is "fed back" as input to the process, closing the loop.

Closed-loop controllers have the following advantages over open-loop controllers:

- Disturbance rejection.
- Guaranteed performance even with model uncertainties, when the model structure does not match perfectly the real process and the model parameters are not exact.

- Unstable processes can be stabilized.
- Reduced sensitivity to parameter variations.
- Improved reference tracking performance.

In some systems, closed-loop and open-loop control are used simultaneously. In such systems, the open-loop control is termed feedforward and serves to further improve reference tracking performance.

A common closed-loop controller architecture is the PID controller.

Classical vs. Modern

A Physical system can be modeled in the "time domain", where the response of a given system is a function of the various inputs, the previous system values, and time. As time progresses, the state of the system and its response change. However, time-domain models for systems are frequently modeled using high-order differential equations which can become impossibly difficult for humans to solve and some of which can even become impossible for modern computer systems to solve efficiently.

To counteract this problem, classical control theory uses the Laplace transform to change an Ordinary Differential Equation (ODE) in the time domain into a regular algebraic polynomial in the frequency domain. Once a given system has been converted into the frequency domain it can be manipulated with greater ease.

Modern control theory, instead of changing domains to avoid the complexities of time-domain ODE mathematics, converts the differential equations into a system of lower-order time domain equations called state equations, which can then be manipulated using techniques from linear algebra.

Laplace Transform

Classical control theory uses the Laplace transform to model the systems and signals. The Laplace transform is a frequency-domain approach for continuous time signals irrespective of whether the system is stable or unstable. The Laplace transform of a function *f*(*t*), defined for all real numbers $t \geq 0$, is the function *F*(*s*), which is a unilateral transform defined by,

$$F(s) = \int_0^\infty e^{-st} f(t)dt$$

where *s* is a complex number frequency parameter

$s = \sigma + i\omega$, with real numbers σ and ω.

Tools

Classical control theory uses an array of tools to analyze systems and design controllers for such systems. Tools include the root locus, the Nyquist stability criterion, the Bode plot, the gain margin and phase margin. More advanced tools include Bode integrals to assess performance limitations and trade-offs, and describing functions to analyze nonlinearities in the frequency domain.

Nonlinear Control

A feedback control system. It is desired to control a system (often called the plant) so its output follows a desired reference signal. A sensor monitors the output and a controller subtracts the actual output from the desired reference output, and applies this error signal to the system to bring the output closer to the reference. In a nonlinear control system at least one of the blocks, system, sensor, or controller, is nonlinear.

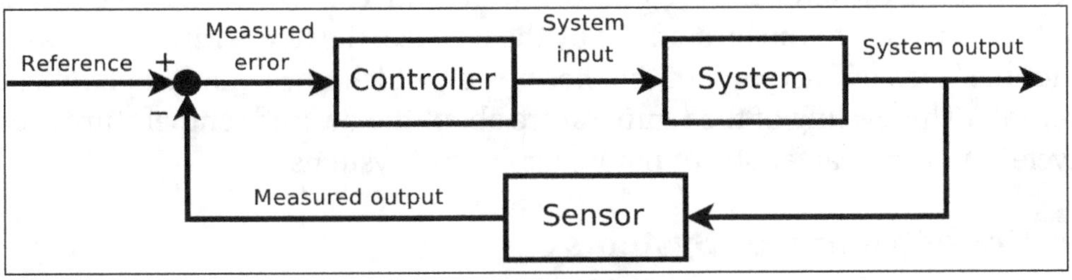

Nonlinear control theory is the area of control theory which deals with systems that are nonlinear, time-variant, or both. Control theory is an interdisciplinary branch of engineering and mathematics that is concerned with the behavior of dynamical systems with inputs, and how to modify the output by changes in the input using feedback, feedforward, or signal filtering. The system to be controlled is called the "plant". One way to make the output of a system follow a desired reference signal is to compare the output of the plant to the desired output, and provide feedback to the plant to modify the output to bring it closer to the desired output.

Control theory is divided into two branches. Linear control theory applies to systems made of devices which obey the superposition principle. They are governed by linear differential equations. A major subclass is systems which in addition have parameters which do not change with time, called linear time invariant (LTI) systems. These systems can be solved by powerful frequency domain mathematical techniques of great generality, such as the Laplace transform, Fourier transform, Z transform, Bode plot, root locus, and Nyquist stability criterion.

Nonlinear control theory covers a wider class of systems that do not obey the superposition principle. It applies to more real-world systems, because all real control systems are nonlinear. These systems are often governed by nonlinear differential equations. The mathematical techniques which have been developed to handle them are more rigorous and much less general, often applying only to narrow categories of systems.

These include limit cycle theory, Poincaré maps, Lyapunov stability theory, and describing functions. If only solutions near a stable point are of interest, nonlinear systems can often be linearized by approximating them by a linear system obtained by expanding the nonlinear solution in a series, and then linear techniques can be used. Nonlinear systems are often analyzed using numerical methods on computers, for example by simulating their operation using a simulation language. Even if the plant is linear, a nonlinear controller can often have attractive features such as simpler implementation, faster speed, more accuracy, or reduced control energy, which justify the more difficult design procedure.

An example of a nonlinear control system is a thermostat-controlled heating system. A building heating system such as a furnace has a nonlinear response to changes in temperature; it is either "on" or "off", it does not have the fine control in response to temperature differences that a proportional (linear) device would have. Therefore, the furnace is off until the temperature falls below the "turn on" setpoint of the thermostat, when it turns on. Due to the heat added by the furnace, the temperature increases until it reaches the "turn off" setpoint of the thermostat, which turns the furnace off, and the cycle repeats. This cycling of the temperature about the desired temperature is called a limit cycle, and is characteristic of nonlinear control systems.

Properties of Nonlinear Systems

Some properties of nonlinear dynamic systems are:

- They do not follow the principle of superposition (linearity and homogeneity).
- They may have multiple isolated equilibrium points.
- They may exhibit properties such as limit cycle, bifurcation, chaos.
- Finite escape time: Solutions of nonlinear systems may not exist for all times.

Analysis and Control of Nonlinear Systems

There are several well-developed techniques for analyzing nonlinear feedback systems:

- Describing function method.
- Phase plane method.
- Lyapunov stability analysis.
- Singular perturbation method.
- The Popov criterion and the circle criterion for absolute stability.
- Center manifold theorem.
- Small-gain theorem.

- Passivity analysis.

Control design techniques for nonlinear systems also exist. These can be subdivided into techniques which attempt to treat the system as a linear system in a limited range of operation and use (well-known) linear design techniques for each region:

- Gain scheduling: Those that attempt to introduce auxiliary nonlinear feedback in such a way that the system can be treated as linear for purposes of control design:
- Feedback linearization: And Lyapunov based methods:
    - Lyapunov redesign.
    - Control-Lyapunov function.
    - Nonlinear damping.
    - Backstepping.
    - Sliding mode control.

Nonlinear Feedback Analysis – The Lur'e Problem

An early nonlinear feedback system analysis problem was formulated by A. I. Lur'e. Control systems described by the Lur'e problem have a forward path that is linear and time-invariant, and a feedback path that contains a memory-less, possibly time-varying, static nonlinearity.

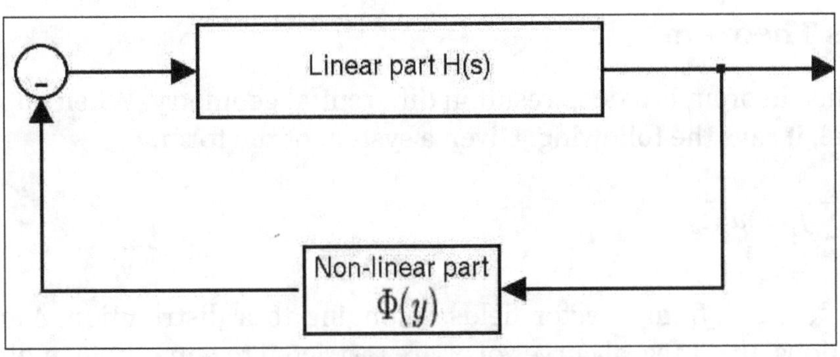

Lur'e problem block diagram.

The linear part can be characterized by four matrices (*A*,*B*,*C*,*D*), while the nonlinear part is $\Phi(y)$ with $\frac{\Phi(y)}{y} \in [a,b], \quad a < b \quad \forall y$ (a sector nonlinearity).

Absolute Stability Problem

Consider:

- (*A*, *B*) is controllable and (*C*, *A*) is observable.

- Two real numbers a, b with $a < b$, defining a sector for function Φ.

The Lur'e problem (also known as the absolute stability problem) is to derive conditions involving only the transfer matrix $H(s)$ and $\{a,b\}$ such that $x = 0$ is a globally uniformly asymptotically stable equilibrium of the system.

There are two well-known wrong conjectures on the absolute stability problem:

- The Aizerman's conjecture,
- The Kalman's conjecture.

Graphically, these conjectures can be interpreted in terms of graphical restrictions on the graph of $\Phi(y)$ x y or also on the graph of $d\Phi/dy$ x Φ/y. There are counterexamples to Aizerman's and Kalman's conjectures such that nonlinearity belongs to the sector of linear stability and unique stable equilibrium coexists with a stable periodic solution—hidden oscillation.

There are two main theorems concerning the Lur'e problem which give sufficient conditions for absolute stability:

- The circle criterion (an extension of the Nyquist stability criterion for linear systems).
- The Popov criterion.

Theoretical Results in Nonlinear Control

Frobenius Theorem

The Frobenius theorem is a deep result in differential geometry. When applied to nonlinear control, it says the following: Given a system of the form,

$$\dot{x} = \sum_{i=1}^{k} f_i(x)u_i(t)$$

where $x \in R^n$, $f_1,\ldots,f_k$ are vector fields belonging to a distribution Δ and $u_i(t)$ are control functions, the integral curves of x are restricted to a manifold of dimension m if $\text{span}(\Delta) = m$ and Δ is an involutive distribution.

Stochastic Control

Stochastic control or stochastic optimal control is a sub field of control theory that deals with the existence of uncertainty either in observations or in the noise that drives the evolution of the system. The system designer assumes, in a Bayesian probability-driven fashion, that random noise with known probability distribution affects the evolution and observation of the state variables. Stochastic control aims to design the time path of the controlled variables that performs the desired control task with minimum cost, somehow defined, despite the presence of this noise. The context may be either discrete time or continuous time.

Certainty Equivalence

An extremely well-studied formulation in stochastic control is that of linear quadratic Gaussian control. Here the model is linear, the objective function is the expected value of a quadratic form, and the disturbances are purely additive. A basic result for discrete-time centralized systems with only additive uncertainty is the certainty equivalence property: that the optimal control solution in this case is the same as would be obtained in the absence of the additive disturbances. This property is applicable to all centralized systems with linear equations of evolution, quadratic cost function, and noise entering the model only additively; the quadratic assumption allows for the optimal control laws, which follow the certainty-equivalence property, to be linear functions of the observations of the controllers.

Any deviation from the above assumptions—a nonlinear state equation, a non-quadratic objective function, noise in the multiplicative parameters of the model, or decentralization of control—causes the certainty equivalence property not to hold. For example, its failure to hold for decentralized control was demonstrated in Witsenhausen's counterexample.

Discrete Time

In a discrete-time context, the decision-maker observes the state variable, possibly with observational noise, in each time period. The objective may be to optimize the sum of expected values of a nonlinear (possibly quadratic) objective function over all the time periods from the present to the final period of concern, or to optimize the value of the objective function as of the final period only. At each time period new observations are made, and the control variables are to be adjusted optimally. Finding the optimal solution for the present time may involve iterating a matrix Riccati equation backwards in time from the last period to the present period.

In the discrete-time case with uncertainty about the parameter values in the transition matrix (giving the effect of current values of the state variables on their own evolution) and/or the control response matrix of the state equation, but still with a linear state equation and quadratic objective function, a Riccati equation can still be obtained for iterating backward to each period's solution even though certainty equivalence does not apply. The discrete-time case of a non-quadratic loss function but only additive disturbances can also be handled, albeit with more complications.

Example:

A typical specification of the discrete-time stochastic linear quadratic control problem is to minimize,

$$E_1 \sum_{t=1}^{S} [y_t^T Q y_t + u_t^T R u_t]$$

where E_1 is the expected value operator conditional on y_0, superscript T indicates a matrix transpose, and S is the time horizon, subject to the state equation,

$$y_t = A_t y_{t-1} + B_t u_t,$$

where y is an $n \times 1$ vector of observable state variables, u is a $k \times 1$ vector of control variables, At is the time t realization of the stochastic $n \times n$ state transition matrix, Bt is the time t realization of the stochastic $n \times k$ matrix of control multipliers, and Q ($n \times n$) and R ($k \times k$) are known symmetric positive definite cost matrices. We assume that each element of A and B is jointly independently and identically distributed through time, so the expected value operations need not be time-conditional.

Induction backwards in time can be used to obtain the optimal control solution at each time,

$$u_t^* = -[\mathrm{E}(B^T X_t B + R)]^{-1} \mathrm{E}(B^T X_t A) y_{t-1},$$

with the symmetric positive definite cost-to-go matrix X evolving backwards in time from $X_S = Q$ according to,

$$X_{t-1} = Q + \mathrm{E}[A^T X_t A] - \mathrm{E}[A^T X_t B][\mathrm{E}(B^T X_t B + R)]^{-1} \mathrm{E}(B^T X_t A),$$

which is known as the discrete-time dynamic Riccati equation of this problem. The only information needed regarding the unknown parameters in the A and B matrices is the expected value and variance of each element of each matrix and the covariances among elements of the same matrix and among elements across matrices.

The optimal control solution is unaffected if zero-mean, i.i.d. additive shocks also appear in the state equation, so long as they are uncorrelated with the parameters in the A and B matrices. But if they are so correlated, then the optimal control solution for each period contains an additional additive constant vector. If an additive constant vector appears in the state equation, then again the optimal control solution for each period contains an additional additive constant vector.

The steady-state characterization of X (if it exists), relevant for the infinite-horizon problem in which S goes to infinity, can be found by iterating the dynamic equation for X repeatedly until it converges; then X is characterized by removing the time subscripts from its dynamic equation.

Continuous Time

If the model is in continuous time, the controller knows the state of the system at each instant of time. The objective is to maximize either an integral of, for example, a concave function of a state variable over a horizon from time zero (the present) to a terminal time T, or a concave function of a state variable at some future date T. As time evolves, new observations are continuously made and the control variables are continuously adjusted in optimal fashion.

Stochastic Model Predictive Control

In the literature, there are two types of MPCs for stochastic systems; Robust model predictive control and Stochastic Model Predictive Control (SMPC). Robust model predictive control is a more conservative method which considers the worst scenario in the optimization procedure. However, this method, similar to other robust controls, deteriorates the overall controller's performance and also is applicable only for systems with bounded uncertainties. The alternative method, SMPC, considers soft constraints which limit the risk of violation by a probabilistic inequality.

In Finance

In a continuous time approach in a finance context, the state variable in the stochastic differential equation is usually wealth or net worth, and the controls are the shares placed at each time in the various assets. Given the asset allocation chosen at any time, the determinants of the change in wealth are usually the stochastic returns to assets and the interest rate on the risk-free asset. The field of stochastic control has developed greatly since the 1970s, particularly in its applications to finance. Robert Merton used stochastic control to study optimal portfolios of safe and risky assets. His work and that of Black–Scholes changed the nature of the finance literature. Influential mathematical textbook treatments were by Fleming and Rishel, and by Fleming and Soner. These techniques were applied by Stein to the financial crisis of 2007–08.

The maximization, say of the expected logarithm of net worth at a terminal date T, is subject to stochastic processes on the components of wealth. In this case, in continuous time Itô's equation is the main tool of analysis. In the case where the maximization is an integral of a concave function of utility over an horizon $(0,T)$, dynamic programming is used. There is no certainty equivalence as in the older literature, because the coefficients of the control variables—that is, the returns received by the chosen shares of assets—are stochastic.

Control System

A control system is a set of mechanical or electronic devices that regulates other devices or systems by way of control loops. Typically, control systems are computerized.

Control systems are a central part of industry and of automation. The types of control loops that regulate these processes include industrial control systems (ICS) such as supervisory control and data acquisition (SCADA) and distributed control systems (DCS).

Basics of Control System

The control system consists of three components:

- Input,
- Logic operation,
- Output or decision device

Input is the cause parameter on which the control system acts. The logic operation is the intended or desired operation to perform on the input for generating a new output state.

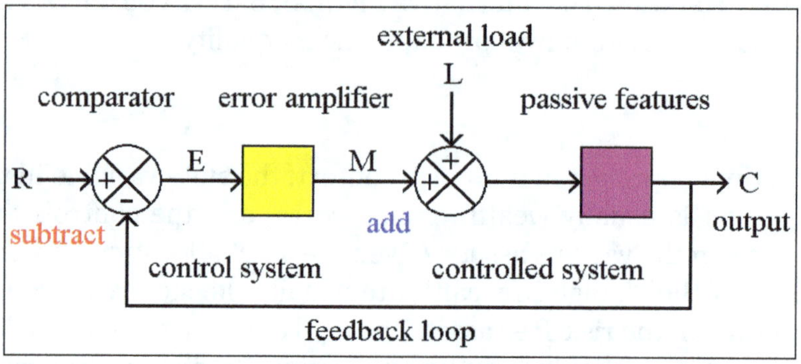

Basic control system.

The output is drive parameter which actuates the end component to perform the desired task. These can be open loop or closed loop control systems depends on output feedback. Block diagram reduction helps in analysis and simplification of control system.

Frequency Domain Analysis

Frequency response of the system is a steady state response the system in which the output may differs in amplitude and phase. Analysis a system in frequency domain is easy compared to time domain analysis. Based on the output response characteristics the behavior of the system can be predicted. The various frequency domain analysis techniques are root locus, bode plots, Nyquist plots etc.

Process Control

Process is a method of changing raw materials in to useful end product. The raw materials changes it state from one form to another form (liquid to gas etc.) during the conversion process. The process industries may be chemical industry, power industry, manufacturing industry. Process control refers to the methods which are used to control the process parameters to obtain a good end product. Now a days increase in electronics growth and enhancement in control strategies almost all the process are automatic to reduce the manual intervention. PID controller and advanced controllers places a major role automatic control system.

Process control.

Control systems are used to enhance production, efficiency and safety in many areas, including:

- Agriculture,
- Chemical plants,
- Pulp and paper mills,
- Quality control,
- Boiler controls and power plant,
- Nuclear power plants,
- Environmental control,
- Water treatment plants,
- Sewage treatment plants,
- Food and food processing,
- Metal and mines,

- Pharmaceutical manufacturing,
- Sugar refining plants.

There are two main types of control loops: Open loops, which operate with human input, and closed loops, which are fully autonomous. Some loops can be switched between closed and open modes. When open, a switchable loop is manually controlled and when closed it is fully automated.

Open-loop System

An Open-loop system, also referred to as non-feedback system, is a type of continuous control system in which the output has no influence or effect on the control action of the input signal. In other words, in an open-loop control system the output is neither measured nor "fed back" for comparison with the input. Therefore, an open-loop system is expected to faithfully follow its input command or set point regardless of the final result. Also, an open-loop system has no knowledge of the output condition so cannot self-correct any errors it could make when the preset value drifts, even if this results in large deviations from the preset value.

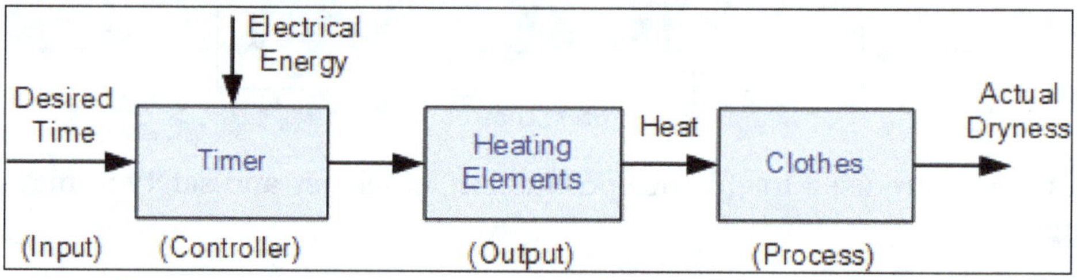

Another disadvantage of open-loop systems is that they are poorly equipped to handle disturbances or changes in the conditions which may reduce its ability to complete the desired task. For example, the dryer door opens and heat is lost. The timing controller continues regardless for the full 30 minutes but the clothes are not heated or dried at the end of the drying process. This is because there is no information fed back to maintain a constant temperature.

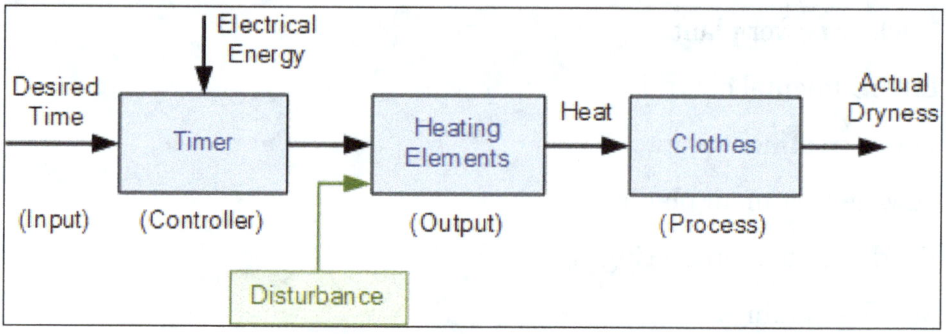

Then we can see that open-loop system errors can disturb the drying process and therefore requires extra supervisory attention of a user (operator). The problem with this anticipatory control approach is that the user would need to look at the process temperature frequently and take any corrective control action whenever the drying process deviated from its desired value of drying the clothes. This type of manual open-loop control which reacts before an error actually occurs is called Feed forward Control.

The objective of feed forward control, also known as predictive control, is to measure or predict any potential open-loop disturbances and compensate for them manually before the controlled variable deviates too far from the original set point. So for our simple example above, if the dryers door was open it would be detected and closed allowing the drying process to continue.

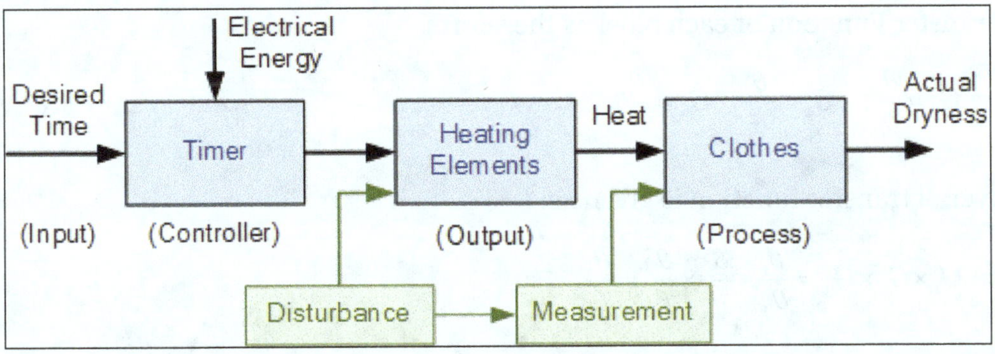

If applied correctly, the deviation from wet clothes to dry clothes at the end of the 30 minutes would be minimal if the user responded to the error situation (door open) very quickly. However, this feed forward approach may not be completely accurate if the system changes, for example the drop in drying temperature was not noticed during the 30 minute process.

Then we can define the main characteristics of an "Open-loop system" as being:

- There is no comparison between actual and desired values.
- An open-loop system has no self-regulation or control action over the output value.
- Each input setting determines a fixed operating position for the controller.
- Changes or disturbances in external conditions does not result in a direct output change (unless the controller setting is altered manually).

Any open-loop system can be represented as multiple cascaded blocks in series or a single block diagram with an input and output. The block diagram of an open-loop system shows that the signal path from input to output represents a linear path with no feedback loop and for any type of control system the input is given the designation θ_i and the output θ_o.

Generally, we do not have to manipulate the open-loop block diagram to calculate its actual transfer function. We can just write down the proper relationships or equations from each block diagram, and then calculate the final transfer function from these equations as shown.

Open-loop System

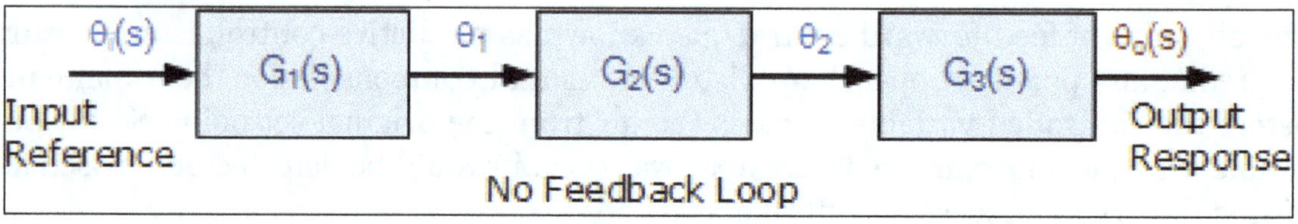

The Transfer Function of each block is therefore:

$$G_1 = \frac{\theta_1}{\theta_i}, \quad G_2 = \frac{\theta_2}{\theta_1}, \quad G_3 = \frac{\theta_o}{\theta_2},$$

The overall transfer function is given as:

$$G_1 \times G_2 \times G_3 = \frac{\cancel{\theta_1}}{\theta_i} \times \frac{\cancel{\theta_2}}{\cancel{\theta_1}} \times \frac{\theta_o}{\cancel{\theta_2}} = \frac{\theta_o}{\theta_i}$$

Then the Open-loop Gain is given simply as:

$$\text{Gain,}(G) = \frac{\theta_o(s)}{\theta_i(s)}$$

When G represents the Transfer Function of the system or subsystem, it can be rewritten as: $G(s) = \theta o(s)/\theta i(s)$.

Open-loop control systems are often used with processes that require the sequencing of events with the aid of "ON-OFF" signals. For example a washing machines which requires the water to be switched "ON" and then when full is switched "OFF" followed by the heater element being switched "ON" to heat the water and then at a suitable temperature is switched "OFF", and so on. This type of "ON-OFF" open-loop control is suitable for systems where the changes in load occur slowly and the process is very slow acting, necessitating infrequent changes to the control action by an operator.

Open-loop Motor Control

So for example, assume the DC motor controller as shown. The speed of rotation of the motor will depend upon the voltage supplied to the amplifier (the controller) by the potentiometer. The value of the input voltage could be proportional to the position of the potentiometer.

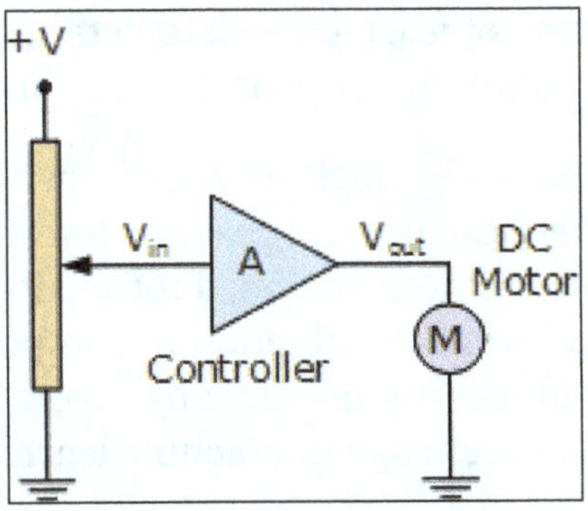

Open-loop motor control.

If the potentiometer is moved to the top of the resistance the maximum positive voltage will be supplied to the amplifier representing full speed. Likewise, if the potentiometer wiper is moved to the bottom of the resistance, zero voltage will be supplied representing a very slow speed or stop. Then the position of the potentiometers slider represents the input, θ_i which is amplified by the amplifier (controller) to drive the DC motor (process) at a set speed N representing the output, θ_o of the system. The motor will continue to rotate at a fixed speed determined by the position of the potentiometer.

As the signal path from the input to the output is a direct path not forming part of any loop, the overall gain of the system will the cascaded values of the individual gains from the potentiometer, amplifier, motor and load. It is clearly desirable that the output speed of the motor should be identical to the position of the potentiometer, giving the overall gain of the system as unity.

However, the individual gains of the potentiometer, amplifier and motor may vary over time with changes in supply voltage or temperature, or the motors load may increase representing external disturbances to the open-loop motor control system.

But the user will eventually become aware of the change in the systems performance (change in motor speed) and may correct it by increasing or decreasing the potentiometers input signal accordingly to maintain the original or desired speed.

The advantages of this type of "open-loop motor control" is that it is potentially cheap and simple to implement making it ideal for use in well-defined systems were the relationship between input and output is direct and not influenced by any outside disturbances. Unfortunately this type of open-loop system is inadequate as variations or disturbances in the system affect the speed of the motor. Then another form of control is required.

Open-loop Controller

In an open-loop controller, also called a non-feedback controller, the control action from the controller is independent of the "process output", which is the process variable that is being controlled. It does not use feedback to determine if its output has achieved the desired goal of the input command or process "set point".

There are many open-loop controls, such as on/off switching of valves, machinery, lights, motors or heaters, where the control result is known to be approximately sufficient under normal conditions without the need for feedback. The advantage of using open-loop control in these cases is the reduction in component count and complexity. However, an open-loop system cannot correct any errors that it makes or correct for outside disturbances, and cannot engage in machine learning.

Open-loop and Closed-loop (Feedback) Control

Fundamentally, there are two types of control loop: open-loop (feedforward) control, and closed loop (feedback) control.

An electromechanical timer, normally used for open-loop control based purely on a timing sequence, with no feedback from the process.

In open-loop control, the control action from the controller is independent of the "process output" (or "controlled process variable"). A good example of this is a central heating boiler controlled only by a timer, so that heat is applied for a constant time, regardless of the temperature of the building. The control action is the switching on/off of the boiler, but the controlled variable should be the building temperature, but is not as this is open-loop control of the boiler, which does not give closed-loop control of the temperature.

In closed loop control, the control action from the controller is dependent on the process output. In the case of the boiler analogy this would include a thermostat to monitor the building temperature, and thereby feedback a signal to ensure the controller maintains the building at the temperature set on the thermostat. A closed loop controller therefore has a feedback loop which ensures the controller exerts a control action to give a process output the same as the "reference input" or "set point". For this reason, closed loop controllers are also called feedback controllers.

The definition of a closed loop control system according to the British Standard Institution is "a control system possessing monitoring feedback, the deviation signal formed as a result of this feedback being used to control the action of a final control element in such a way as to tend to reduce the deviation to zero."

Applications

Electric clothes dryer, which is open-loop controlled by running the dryer for a set time, regardless of clothes dryness.

An open-loop controller is often used in simple processes because of its simplicity and low cost, especially in systems where feedback is not critical. A typical example would be an older model domestic clothes dryer, for which the length of time is entirely dependent on the judgement of the human operator, with no automatic feedback of the dryness of the clothes.

For example, an irrigation sprinkler system, programmed to turn on at set times could be an example of an open-loop system if it does not measure soil moisture as a form of feedback. Even if rain is pouring down on the lawn, the sprinkler system would activate on schedule, wasting water.

Another example is a Stepper motors used for control of position. Sending it a stream of electrical pulses causes it to rotate by exactly that many steps, hence the name. If the motor was always assumed to perform each movement correctly, without positional feedback, it would be open-loop control. However, if there is a position encoder, or sensors to indicate the "start" or finish positions, then that is closed-loop control, such as in many inkjet printers. The drawback of open-loop control of steppers is that if the machine load is too high, or the motor attempts to move too quickly, then steps may be skipped. The controller has no means of detecting this and so the machine continues to run slightly out of adjustment until reset. For this reason, more complex robots and machine tools instead use servomotors rather than stepper motors, which incorporate encoders and closed-loop controllers.

However, open-loop control is very useful and economic for well-defined systems where the relationship between input and the resultant state can be reliably modeled by a mathematical formula. For example, determining the voltage to be fed to an electric motor that drives a constant load, in order to achieve a desired speed would be a good application. But if the load were not predictable and became excessive, the motor's speed might vary as a function of the load not just the voltage, and an open-loop controller would be insufficient to ensure repeatable control of the velocity.

An example of this is a conveyor system that is required to travel at a constant speed. For a constant voltage, the conveyor will move at a different speed depending on the load on the motor (represented here by the weight of objects on the conveyor). In order for the conveyor to run at a constant speed, the voltage of the motor must be adjusted depending on the load. In this case, a closed-loop control system would be necessary.

Thus there are many open-loop controls, such as switching valves, lights, motors or heaters on and off, where the result is known to be approximately sufficient without the need for feedback.

Feedback Control

A feedback control system, such as a PID controller, can be improved by combining the feedback (or closed-loop) control of a PID controller with feed-forward (or open-loop) control. Knowledge about the system (such as the desired acceleration and inertia) can be fed forward and combined with the PID output to improve the overall system performance. The feed-forward value alone can often provide the major portion of the controller output. The PID controller primarily has to compensate whatever difference or error remains between the setpoint (SP) and the system response to the open-loop control. Since the feed-forward output is not affected by the process feedback, it can never cause the control system to oscillate, thus improving the system response without affecting stability. Feed forward can be based on the setpoint and on extra measured disturbances. Setpoint weighting is a simple form of feed forward.

For example, in most motion control systems, in order to accelerate a mechanical load under control, more force is required from the actuator. If a velocity loop PID controller is being used to control the speed of the load and command the force being applied by the actuator, then it is beneficial to take the desired instantaneous acceleration, scale that value appropriately and add it to the output of the PID velocity loop controller. This means that whenever the load is being accelerated or decelerated, a proportional amount of force is commanded from the actuator regardless of the feedback value. The PID loop in this situation uses the feedback information to change the combined output to reduce the remaining difference between the process setpoint and the feedback value. Working together, the combined open-loop feed-forward controller and closed-loop PID controller can provide a more responsive control system in some situations.

Closed-loop System

A control system with feedback loop is called "closed loop control system". In other words, the control system which uses its feedback signal to generate output is called "closed loop control system". In these control systems, the input is controlled by the feedback signal from input so that it can correct the errors occurred. Closed loop control systems are two way signal flow systems.

Feedback means some part of output is taken and connected it to the input of the system to maintain the stability of the control system. By providing a feedback loop, we can convert any open loop control system into closed loop system. The feedback loop provides the automatic correction of the input signal based on the output requirement.

By comparing the generated output with the actual condition, the closed loop system maintains and achieves the desired output. If the produced output is deviated from decided (actual) output, the closed loop control system generates an error signal and the error signal is fed to the input of the signal. So by adding the error signal to the input, the generated output of the next loop will be corrected. So these are also called as automatic control systems. Closed loop systems are less prone to external disturbances.

The block diagram of the closed loop system is shown below:

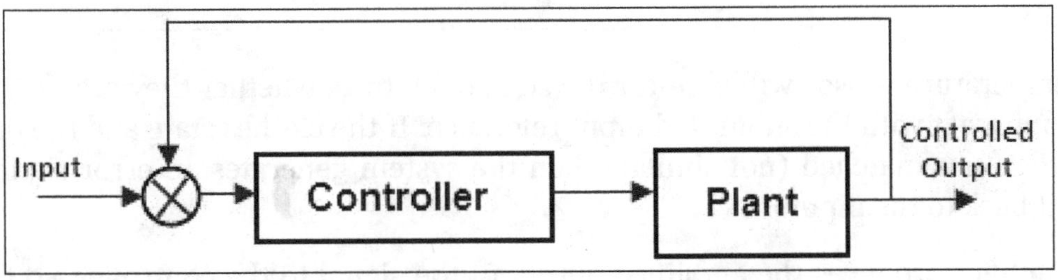

As these systems contain feedback loop, the closed loop control systems are also called as "Feedback control systems". By giving the feedback to the input signal, we can accurately control the output of a control system. A closed loop control system can have more than one feedback loop.

Examples: We use closed loop control systems in many applications of our day to day life. Some of the systems designed based on the concept of open loop control systems are given below:

- Automatic Electric Iron: Depending on the temperature of the iron heating elements were controlled automatically.

- Servo Voltage Stabilizer: Stabilization in voltage is achieved by the feeding the output voltage back to the system.
- Water Level Controller: Water level in the reservoirs decides the input water into it.
- Air Conditioner: Air conditioner automatically adjusts its temperature depending on its room temperature.
- In motor speed regulator using tachometer and/or current sensor, sensor senses the speed and sends a feedback to the system to regulate its speed.

Practical Example of Feedback System

In market we see some automatic cloth driers. Does anybody know that they are an example of closed loop control system? In automatic control system or closed loop feedback system, we use a transducer at input side of the system to continuously monitor the temperature of the clothes inside the drier. The block diagram of the electric cloth drier is shown below:

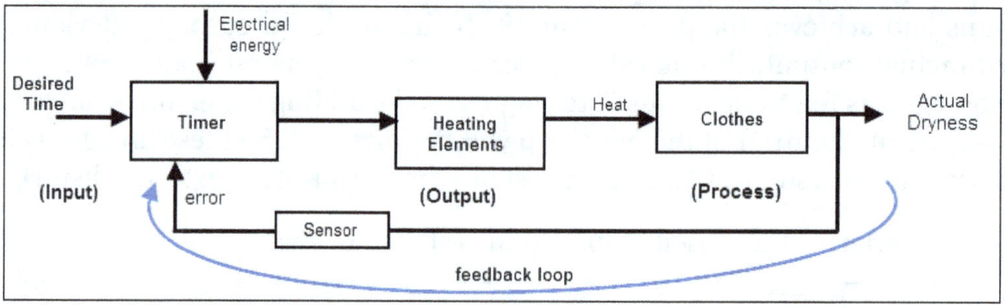

The temperature sensor will monitor the state of clothes, whether they are dry or wet and compare it with the submitted input reference. If the clothes state and the desired output are not matched (not similar) then the system generates an error signal and sends it back to the input.

The amplifier produces the amplified output of the signal to the controller, so that, it can adjust the heating elements to the desired output level. So if the clothes are about to dry, then the controller regulates and lowers the temperature or stops the heating process of the heating elements to an extent that it doesn't burn the clothes in the machine.

The closed loop system is capable of handling the external noise disturbances (which causes to reduce the system efficiency) by means of the feedback signal. For example, if we open the drier, some amount of temperature is lost and the inner temperature of the system will decrease. Then the feedback sensor will calculate the temperature of the clothes and then it sends the error signal to controller. So that the controller sets the required temperature in the system. This will be the process of handling the error occurrence in clothes drier.

The error signal means the difference between actual output and the produced output and this error signal is provided as an input to the controller to regulate the system output. If the error signal is zero, then only the desired input is equal to the produced output.

Types of Feedback Systems

Based on the nature of feedback signal the closed loop control systems are divided in to 2 types. They are:

- Positive feedback signal,
- Negative feedback signal.

Positive Feedback Signal

The closed loop system with positive feedback signal connected to its input is called "Positive feedback system". The block diagram of positive feedback system is shown below. Positive feedback also known as "Regenerative feedback".

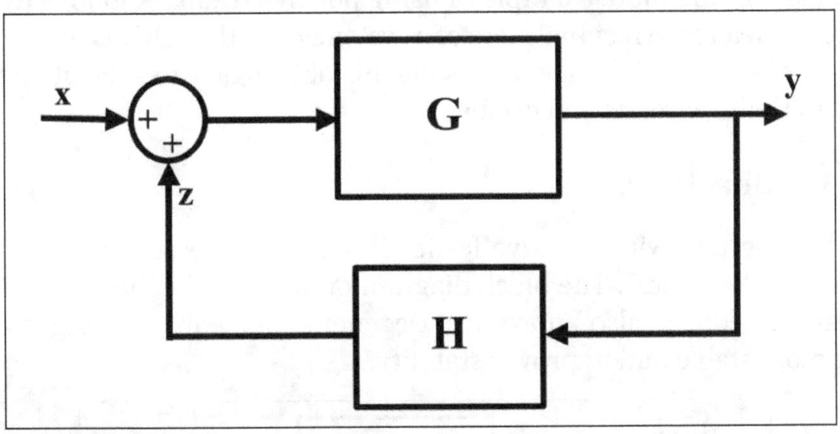

Example: The best example for positive feedback loop in electronics is 'op-amp'. We can achieve the positive feedback loop by connecting some part of the output voltage to the non-inverting input by a feedback loop, through a resistor.

If we connect a positive input voltage signal to the non-inverting port of the op-amp, it amplifies the input signal and the output will become more positive. So some part of the output is connected back to the input by the feedback loop to produce even high output voltage. As the feedback loop is connected to the input, the output becomes more and more positive. At a point, the output of the op-amp will become saturated. Similarly if we supply a negative input voltage signal V_{in} to the op-amp, then the output will become negative. The feedback loop connected to the input will strengthen the amplified signal to become more negative. This means, the feedback signal in positive feedback systems will strengthen the nature of output signal.

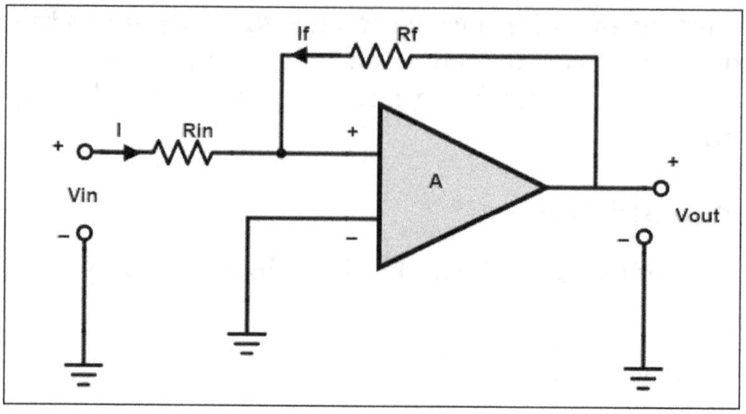

The transfer function of positive loop gained system will be A¬V = G / (1-GH). If GH = 1, then the gain of the system will be infinity, thus the system will automatically oscillate even though no input is connected to it. At this state, the system will act like an oscillator.

The positive feedback systems are used to obtain a fast switching response for a conditional input signal. One more example of using positive feedback loop is hysteresis effect. Hysteresis means the residual state of a system, even though it is not supplied with any input. We observe such behavior in some bistable logic gates. Positive feedback is widely used in timing circuits and oscillators.

Negative Feedback Signal

The closed loop system with negative feedback signal connected to its input is called "Negative feedback system". The block diagram of negative feedback system is shown below. Negative feedback also known as "Degenerative feedback". Negative feedback systems are more stable and improves stability.

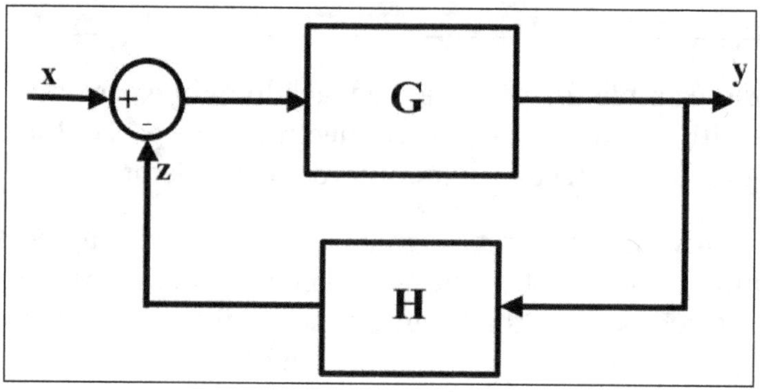

Applications

The closed loop control systems are used to control the electronic machines such as voltage and current generators. Apart from that they can also control the speed of a machinery and other parameters of a system. The closed loop control systems are used for the below stated requirements:

- Enhancement of speed of response,
- In Protection of devices,
- To improve accuracy.

Closed Loop Torque Control

The torque control technique will be used in battery operated vehicles such as bikes and cars. When the driver pressed the accelerator, the reference torque T is set by driver. If the torque controlled by the driver is T* then the actual torque T will follow the controlled torque.

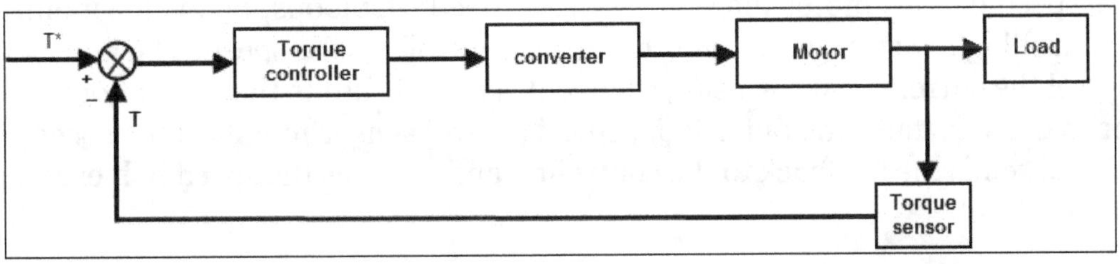

The torque sensor placed at the load will send the feedback about motor speed and helps in maintaining the required speed.

Current Limit Control

In electronic devices, there will be a large amount of current flow by system faults and connection faults. If we doesn't control the current and unless make some precautionary actions, the electronic device will damage and lead to loss, economically and industrially. So to limit the current and to control the current flow, we instantiate a current limit controller in the electronic device or system. The current sensor will continuously monitor the current supplied to the motor.

When the current flow in the motor exceeds the required safe level the sensor activates the feedback loop and some amount of current is negatively feed backed to the controller to set the current back to the limit.

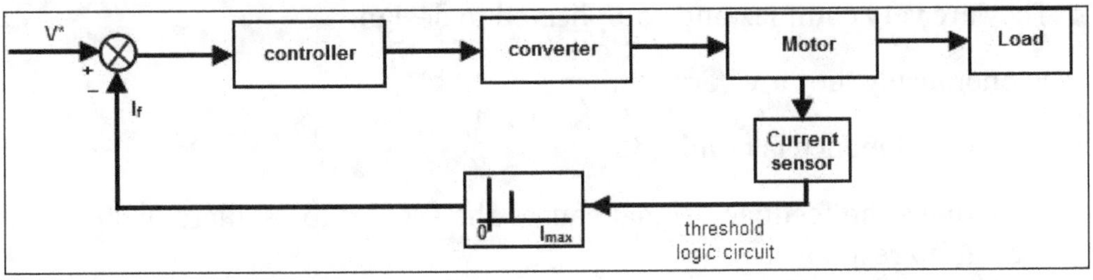

When the current flow again reaches the safe limit, the feedback loop deactivates.

Closed Loop Speed Control

This is the most used feedback loop for the drivers. Observe the diagram below, which uses the current sensor and speed sensor to limit the speed of the motor.

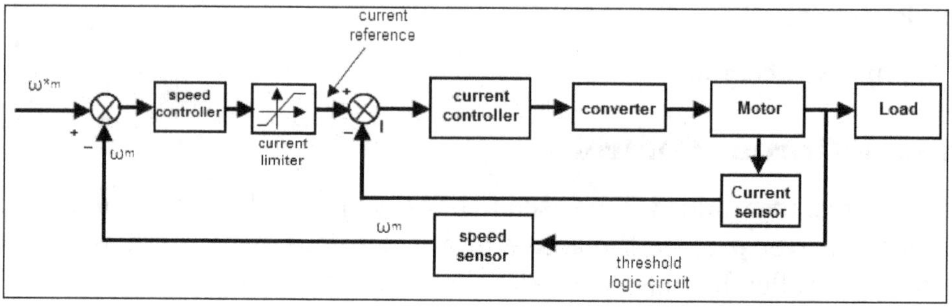

In this speed control motor diagram, we see 2 feedback loops, one is for monitoring and controlling the current and another is for controlling the speed. The inner loop is to control the current flow, by using a current sensor. It limits the current or torque of motor to the safe imit. The outer loop, speed control sensor monitors the speed of the motor and sends the feedback to the controller, and controls the speed in it exceeds the predefined level.

Advantages of Open Loop Control System

- As the closed loop control systems have feedback signal to control the output, these are very accurate and less error prone.
- They can automatically correct the errors by means of the feedback signal.
- Closed loop systems are very accurate.
- Band width of open loop system is very large than closed loop system.
- They can support automation.
- As they have very high noise margin, they are less affected by noise.

Disadvantages of Closed Loop Control System

- They are very complex and complicated to design.
- Economically they are very costlier.
- High maintenance is required.
- Sometimes the feedback signal causes the system to oscillate, which will give oscillatory response.
- More time and efforts are needed to design a stable closed loop system

Closed-loop Transfer Function

A closed-loop transfer function in control theory is a mathematical expression (algorithm) describing the net result of the effects of a closed (feedback) loop on the input signal to the circuits enclosed by the loop.

The closed-loop transfer function is measured at the output. The output signal waveform can be calculated from the closed-loop transfer function and the input signal waveform.

An example of a closed-loop transfer function is shown below:

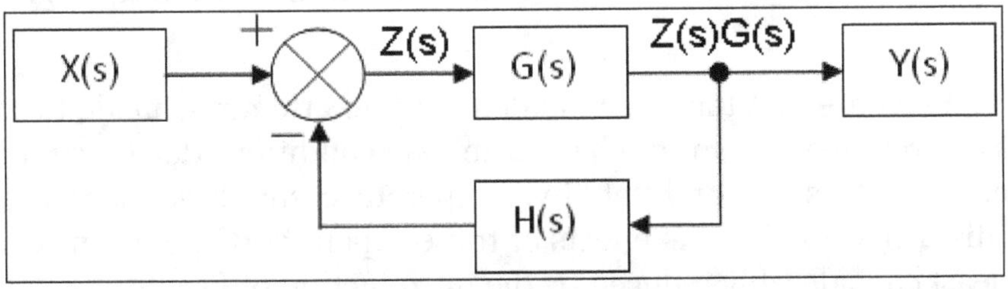

The summing node and the *G*(*s*) and *H*(*s*) blocks can all be combined into one block, which would have the following transfer function:

$$\frac{Y(s)}{X(s)} = \frac{G(s)}{1+G(s)H(s)}$$

Derivation

We define an intermediate signal Z shown as follows:

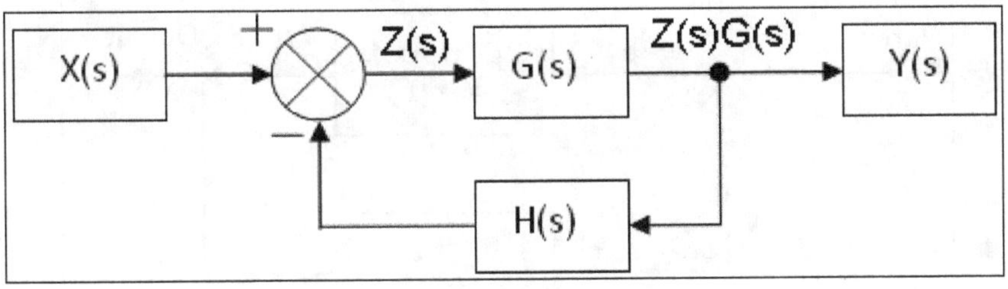

Using this figure we write:

$$Y(s) = G(s)Z(s)$$
$$Z(s) = X(s) - H(s)Y(s)$$
$$Y(s) = G(s)(X(s) - H(s)Y(s)) = G(s)X(s) - G(s)H(s)Y(s)$$
$$Y(s) + G(s)H(s)Y(s) = G(s)X(s)$$
$$Y(s)(1+G(s)H(s)) = G(s)X(s)$$
$$\Rightarrow \frac{Y(s)}{X(s)} = \frac{G(s)}{1+G(s)H(s)}$$

Feedback Systems

In a Feedback System, all or part of the output signal either positive or negative is fed back to the input. Feedback Systems process signals and as such are signal processors. The processing part of a feedback system may be electrical or electronic, ranging from a very simple to a highly complex circuits.

Simple analogue feedback control circuits can be constructed using individual or discrete components, such as transistors, resistors and capacitors, etc. or by using microprocessor-based and integrated circuits (IC's) to form more complex digital feedback systems.

Open-loop systems are just that, open ended, and no attempt is made to compensate for changes in circuit conditions or changes in load conditions due to variations in circuit parameters, such as gain and stability, temperature, supply voltage variations and/or external disturbances. But the effects of these "open-loop" variations can be eliminated or at least considerably reduced by the introduction of Feedback.

A feedback system is one in which the output signal is sampled and then fed back to the input to form an error signal that drives the system. Feedback Systems are very useful and widely used in amplifier circuits, oscillators, process control systems as well as other types of electronic systems. But for feedback to be an effective tool it must be controlled as an uncontrolled system will either oscillate or fail to function. The basic model of a feedback system is given as:

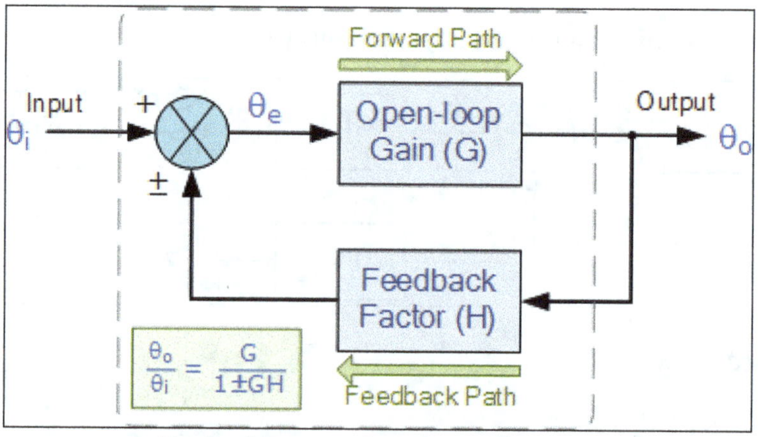

Feedback System Block Diagram Model.

This basic feedback loop of sensing, controlling and actuation is the main concept behind a feedback control system and there are several good reasons why feedback is applied and used in electronic circuits:

- Circuit characteristics such as the systems gain and response can be precisely controlled.

- Circuit characteristics can be made independent of operating conditions such as supply voltages or temperature variations.
- Signal distortion due to the non-linear nature of the components used can be greatly reduced.
- The Frequency Response, Gain and Bandwidth of a circuit or system can be easily controlled to within tight limits.

Whilst there are many different types of control systems, there are just two main types of feedback control namely: Negative Feedback and Positive Feedback.

Positive Feedback Systems

In a "positive feedback control system", the set point and output values are added together by the controller as the feedback is "in-phase" with the input. The effect of positive (or regenerative) feedback is to "increase" the systems gain, i.e. the overall gain with positive feedback applied will be greater than the gain without feedback. For example, if someone praises you or gives you positive feedback about something, you feel happy about yourself and are full of energy, you feel more positive.

However, in electronic and control systems to much praise and positive feedback can increase the systems gain far too much which would give rise to oscillatory circuit responses as it increases the magnitude of the effective input signal. An example of a positive feedback systems could be an electronic amplifier based on an operational amplifier, or op-amp as shown.

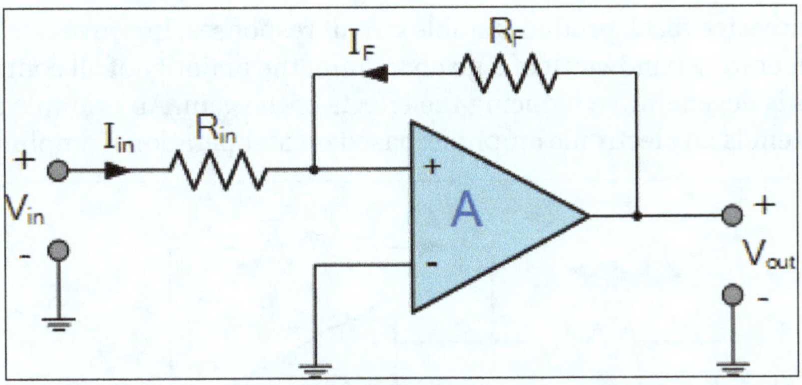

Positive Feedback System.

Positive feedback control of the op-amp is achieved by applying a small part of the output voltage signal at V_{out} back to the non-inverting (+) input terminal via the feedback resistor, R_F.

If the input voltage V_{in} is positive, the op-amp amplifies this positive signal and the output becomes more positive. Some of this output voltage is returned back to the input by the feedback network. Thus the input voltage becomes more positive, causing an even larger output voltage and so on. Eventually the output becomes saturated at its positive supply rail.

Likewise, if the input voltage V_{in} is negative, the reverse happens and the op-amp saturates at its negative supply rail. Positive feedback does not allow the circuit to function as an amplifier as the output voltage quickly saturates to one supply rail or the other, because with positive feedback loops "more leads to more" and "less leads to less".

Then if the loop gain is positive for any system the transfer function will be: $Av = G / (1 - GH)$. If $GH = 1$ the system gain Av = infinity and the circuit will start to self-oscillate, after which no input signal is needed to maintain oscillations, which is useful if you want to make an oscillator.

Although often considered undesirable, this behaviour is used in electronics to obtain a very fast switching response to a condition or signal. One example of the use of positive feedback is hysteresis in which a logic device or system maintains a given state until some input crosses a preset threshold. This type of behaviour is called "bi-stability" and is often associated with logic gates and digital switching devices such as multivibrators.

Positive or regenerative feedback increases the gain and the possibility of instability in a system which may lead to self-oscillation and as such, positive feedback is widely used in oscillatory circuits such as Oscillators and Timing circuits.

In a "negative feedback control system", the set point and output values are subtracted from each other as the feedback is "out-of-phase" with the original input. The effect of negative (or degenerative) feedback is to "reduce" the gain.

Because negative feedback produces stable circuit responses, improves stability and increases the operating bandwidth of a given system, the majority of all control and feedback systems is degenerative reducing the effects of the gain. An example of a negative feedback system is an electronic amplifier based on an operational amplifier as shown.

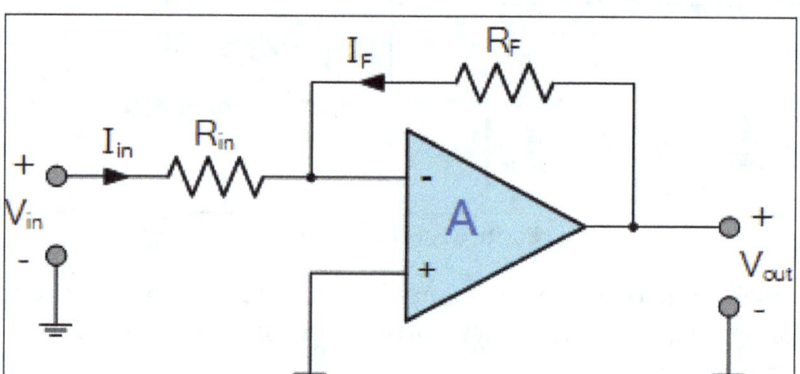

Negative Feedback System.

Negative feedback control of the amplifier is achieved by applying a small part of the output voltage signal at V_{out} back to the inverting (–) input terminal via the feedback resistor, R_f.

If the input voltage V_{in} is positive, the op-amp amplifies this positive signal, but because its connected to the inverting input of the amplifier, and the output becomes more negative. Some of this output voltage is returned back to the input by the feedback network of R_f. Thus the input voltage is reduced by the negative feedback signal, causing an even smaller output voltage and so on. Eventually the output will settle down and become stabilised at a value determined by the gain ratio of $R_f \div R_{in}$.

Likewise, if the input voltage V_{in} is negative, the reverse happens and the op-amps output becomes positive (inverted) which adds to the negative input signal. Then we can see that negative feedback allows the circuit to function as an amplifier, so long as the output is within the saturation limits. The output voltage is stabilised and controlled by the feedback, because with negative feedback loops "more leads to less" and "less leads to more".

Then if the loop gain is positive for any system the transfer function will be: $Av = G / (1 + GH)$.

The use of negative feedback in amplifier and process control systems is widespread because as a rule negative feedback systems are more stable than positive feedback systems, and a negative feedback system is said to be stable if it does not oscillate by itself at any frequency except for a given circuit condition. Another advantage is that negative feedback also makes control systems more immune to random variations in component values and inputs. It must be used with caution as negative feedback significantly modifies the operating characteristics of a given system.

Classification of Feedback Systems

Based on the input quantity being amplified, and on the desired output condition, the input and output variables can be modelled as either a voltage or a current. As a result, there are four basic classifications of single-loop feedback system in which the output signal is fed back to the input and these are:

- Series-Shunt Configuration: Voltage in and Voltage out or Voltage Controlled Voltage Source (VCVS).
- Shunt-Shunt Configuration: Current in and Voltage out or Current Controlled Voltage Source (CCVS).
- Series-Series Configuration: Voltage in and Current out or Voltage Controlled Current Source (VCCS).
- Shunt-Series Configuration: Current in and Current out or Current Controlled Current Source (CCCS).

These names come from the way that the feedback network connects between the input and output stages as shown.

Series-Shunt Feedback Systems

Series-Shunt Feedback, also known as series voltage feedback, operates as a voltage-voltage controlled feedback system. The error voltage fed back from the feedback network is in series with the input. The voltage which is fed back from the output being proportional to the output voltage, Vo as it is parallel, or shunt connected.

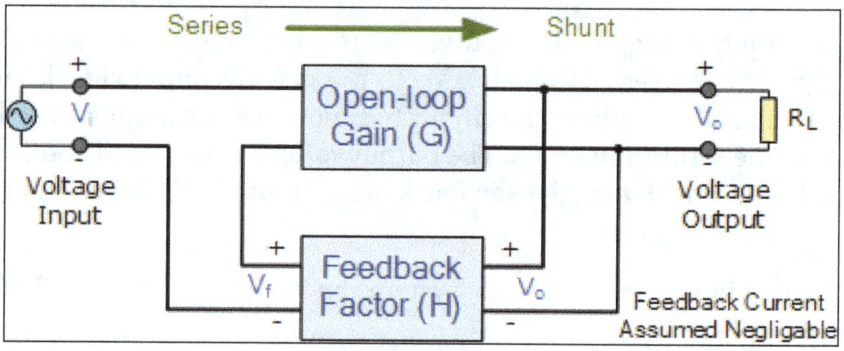

Series-Shunt Feedback System.

For the series-shunt connection, the configuration is defined as the output voltage, V_{out} to the input voltage, V_{in}. Most inverting and non-inverting operational amplifier circuits operate with series-shunt feedback producing what is known as a "voltage amplifier". As a voltage amplifier the ideal input resistance, R_{in} is very large, and the ideal output resistance, R_{out} is very small.

Then the "series-shunt feedback configuration" works as a true voltage amplifier as the input signal is a voltage and the output signal is a voltage, so the transfer gain is given as: $Av = V_{out} \div V_{in}$.

Shunt-Series Feedback Systems

Shunt-Series Feedback, also known as shunt current feedback, operates as a current-current controlled feedback system. The feedback signal is proportional to the output current, Io flowing in the load. The feedback signal is fed back in parallel or shunt with the input.

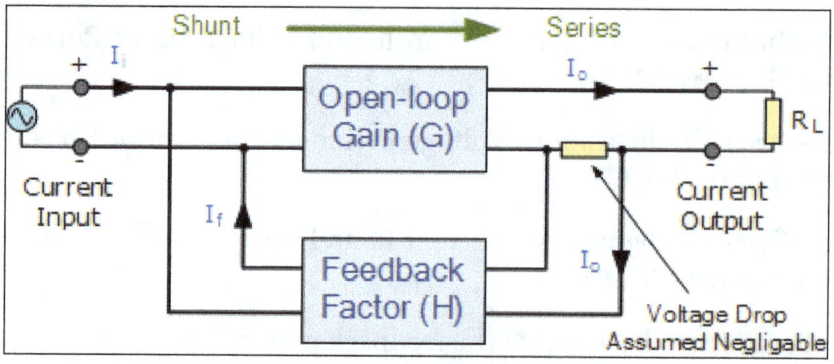

Shunt-Series Feedback System.

For the shunt-series connection, the configuration is defined as the output current, I_{out} to the input current, I_{in}. In the shunt-series feedback configuration the signal fed back is in parallel with the input signal and as such as the currents, not the voltages that add.

This parallel shunt feedback connection will not normally affect the voltage gain of the system, since for a voltage output a voltage input is required. Also, the series connection at the output increases output resistance, R_{out} while the shunt connection at the input decreases the input resistance, R_{in}.

Then the "shunt-series feedback configuration" works as a true current amplifier as the input signal is a current and the output signal is a current, so the transfer gain is given as: $A_i = I_{out} \div I_{in}$. This quantity is dimensionless as its units are amperes/amperes.

Series-Series Feedback Systems

Series-Series Feedback Systems, also known as series current feedback, operates as a voltage-current controlled feedback system. In the series current configuration the feedback error signal is in series with the input and is proportional to the load current, I_{out}. Actually, this type of feedback converts the current signal into a voltage which is actually fed back and it is this voltage which is subtracted from the input.

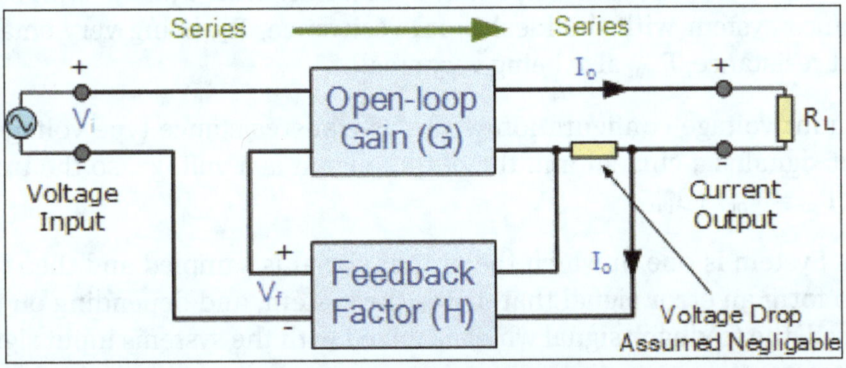

Series-Series Feedback System.

For the series-series connection, the configuration is defined as the output current, I_{out} to the input voltage, V_{in}. Because the output current, I_{out} of the series connection is fed back as a voltage, this increases both the input and output impedances of the system. Therefore, the circuit works best as a transconductance amplifier with the ideal input resistance, R_{in} being very large, and the ideal output resistance, R_{out} is also very large.

Then the "series-series feedback configuration" functions as transconductance type amplifier system as the input signal is a voltage and the output signal is a current. Then for a series-series feedback circuit the transfer gain is given as: $G_m = I_{out} \div V_{in}$.

Shunt-Shunt Feedback Systems

Shunt-Shunt Feedback Systems, also known as shunt voltage feedback, operates as a current-voltage controlled feedback system. In the shunt-shunt feedback configuration the signal fed back is in parallel with the input signal. The output voltage is sensed and the current is subtracted from the input current in shunt, and as such as the currents, not the voltages that subtract.

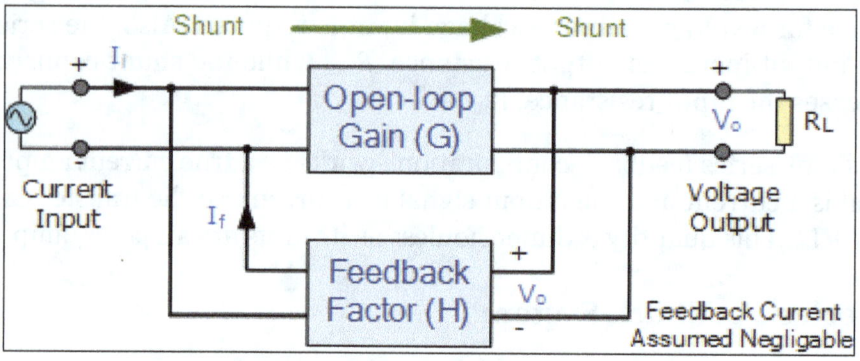

Shunt-Shunt Feedback System.

For the shunt-shunt connection, the configuration is defined as the output voltage, V_{out} to the input current, I_{in}. As the output voltage is fed back as a current to a current-driven input port, the shunt connections at both the input and output terminals reduce the input and output impedance. therefore the system works best as a transresistance system with the ideal input resistance, R_{in} being very small, and the ideal output resistance, R_{out} also being very small.

Then the shunt voltage configuration works as transresistance type voltage amplifier as the input signal is a current and the output signal is a voltage, so the transfer gain is given as: $R_m = V_{out} \div I_{in}$.

A Feedback System is one in which the output signal is sampled and then fed back to the input to form an error signal that drives the system, and depending on the type of feedback used, the feedback signal which is mixed with the systems input signal, can be either a voltage or a current.

Feedback will always change the performance of a system and feedback arrangements can be either positive (regenerative) or negative (degenerative) type feedback systems. If the feedback loop around the system produces a loop-gain which is negative, the feedback is said to be negative or degenerative with the main effect of the negative feedback is in reducing the systems gain.

If however the gain around the loop is positive, the system is said to have positive feedback or regenerative feedback. The effect of positive feedback is to increase the gain which can cause a system to become unstable and oscillate especially if GH = -1.

Block-diagrams can be used to demonstrate the various types of feedback systems. In the block diagrams above, the input and output variables can be modelled as either a voltage or a current and as such there are four combinations of inputs and outputs that represent the possible types of feedback, namely: Series Voltage Feedback, Shunt Voltage Feedback, Series Current Feedback and Shunt Current Feedback.

The names for these different types of feedback systems are derived from the way that the feedback network connects between the input and output stages either in parallel (shunt) or series.

Negative Feedback Systems

- Negative Feedback is the most common form of feedback control configuration used in process, micro-computer and amplification systems.
- Feedback is the process by which a fraction of the output signal, either a voltage or a current, is used as an input. If this feedback fraction is opposite in value or phase ("anti-phase") to the input signal, then the feedback is said to be Negative Feedback, or degenerative feedback.
- Negative feedback opposes or subtracts from the input signals giving it many advantages in the design and stabilisation of control systems. For example, if the systems output changes for any reason, then negative feedback affects the input in such a way as to counteract the change.
- Feedback reduces the overall gain of a system with the degree of reduction being related to the systems open-loop gain. Negative feedback also has effects of reducing distortion, noise, sensitivity to external changes as well as improving system bandwidth and input and output impedances.

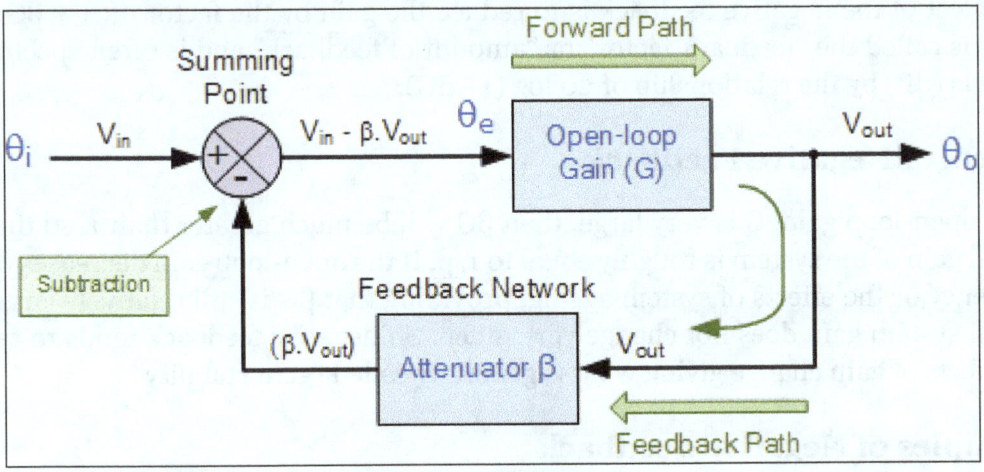

Negative Feedback Circuit.

- Feedback in an electronic system, whether negative feedback or positive feedback is unilateral in direction. Meaning that its signals flow one way only from the output to the input of the system. This then makes the loop gain, G of the system independent of the load and source impedances. As feedback implies a closed-loop system it must therefore have a summing point. In a negative feedback system this summing point or junction at its input subtracts the feedback signal from the input signal to form an error signal, β which drives the system. If the system has a positive gain, the feedback signal must be subtracted from the input signal in order for the feedback to be negative.

The circuit represents a system with positive gain, G and feedback, β. The summing junction at its input subtracts the feedback signal from the input signal to form the error signal V_{in} - βG, which drives the system.

Then using the basic closed-loop circuit above we can derive the general feedback equation as being:

Negative Feedback Equation

$$\text{System Gain, } G = \frac{V_{out}}{V_{in}} \quad G = \text{open loop Voltage gain}$$

$$G \times V_{in}\ V_{out}$$

$$G(V_{in} - \beta V_{out}) = V_{out} \quad \beta \text{ is the feedback fraction}$$

$$G.V_{in} - \beta.G.V_{out} = V_{out} \quad \beta G = \text{the loop gain}$$

$$G.V_{in} = V_{out}(1+\beta G) \quad 1+\beta G = \text{the feedback factor}$$

$$\therefore \frac{V_{out}}{V_{in}} = Gv = \frac{G}{1+\beta G} \quad GV = \text{closed loop voltage gain}$$

The effect of the negative feedback is to reduce the gain by the factor of: 1 + βG. This factor is called the "feedback factor" or "amount of feedback" and is often specified in decibels (dB) by the relationship of 20 log (1+ βG).

Effects of Negative Feedback

If the open-loop gain, G is very large, then βG will be much greater than 1, so that the overall gain of the system is roughly equal to 1/β. If the open-loop gain decreases due to frequency or the effects of system ageing, providing that βG is still relatively large, the overall system gain does not change very much. So negative feedback tends to reduce the effects of gain change giving what is generally called "gain stability".

Examples of Negative Feedback

Example: A system has a gain of 80dB without feedback. If the negative feedback fraction is 1/50th. Calculate the closed-loop gain of the system in dB with the addition of negative feedback.

$$80\,\text{dB} = 20\log(G)$$

$$\therefore G = \text{antilog}\,10^{80/20} = 10{,}000$$

$$Gv = \frac{G}{1+G\beta} = \frac{10{,}000}{1+10{,}000.\left(\frac{1}{50}\right)} = 49.75$$

$$\therefore Gv_{(dB)} = 20\log(49.75) = 34\,\text{dB}$$

Then we can see that the system has a loop gain of 10,000 and a closed-loop gain of 34dB.

Example: If after 5 years the loop gain of the system without negative feedback has fallen to 60dB and the feedback fraction has remained constant at 1/50th. Calculate the new closed-loop gain value of the system.

$$60\,\text{dB} = 20\log(G)$$

$$\therefore G = \text{antilog}\,10^{60/20} = 1{,}000$$

$$Gv = \frac{G}{1+G\beta} = \frac{1{,}000}{1+1{,}000.\left(\frac{1}{50}\right)} = 47.6$$

$$\therefore Gv_{(dB)} = 20\log(47.6) = 33.5\,\text{dB}$$

Then we can see from the two examples that without feedback, after 5 years of use the systems gain has fallen from 80dB down to 60dB, (10,000 to 1,000) a drop in open loop gain of about 25%. However with the addition of negative feedback the systems gain has only fallen from 34dB to 33.5dB, a reduction of less than 1.5%, which proves that negative feedback gives added stability to a systems gain. Therefore we can see that by applying negative feedback to a system greatly reduces its overall gain compared to its gain without feedback.

The systems gain without feedback can be very large but not precise as it may change from one system device to the next, then it is possible to design a system with sufficient open-loop gain that, after the negative feedback has been added, the overall gain matches the desired value.

Also, if the feedback network is constructed from passive elements having stable characteristics, the overall gain becomes very steady and unaffected by variation in the systems inherent open-loop gain.

Negative Feedback in Operational Amplifiers

Operational amplifiers (op-amps) are the most commonly used type of linear integrated circuit but they have a very high gain. The open-loop voltage gain, A_{VOL}, of a standard 741 op-amp is its voltage gain when there is no negative feedback applied and the open-loop voltage gain of an op-amp is the ratio of its output voltage, V_{out}, to its differential input voltage, V_{in}, (V_{out}/V_{in}).

The typical value of A_{VOL} for a 741 op-amp is more than 200,000 (106dB). So an input voltage signal of only 1mV, would result in an output voltage of over 200 volts! forcing the output immediately into saturation. Obviously this high open-loop voltage gain needs to be controlled in some way, and we can do just that by using negative feedback.

The use of negative feedback can significantly improve the performance of an operational amplifier and any op-amp circuit that does not use negative feedback is considered too unstable to be useful. But how can we use negative feedback to control an op-amp. Well consider the circuit below of a Non-inverting Operational Amplifier.

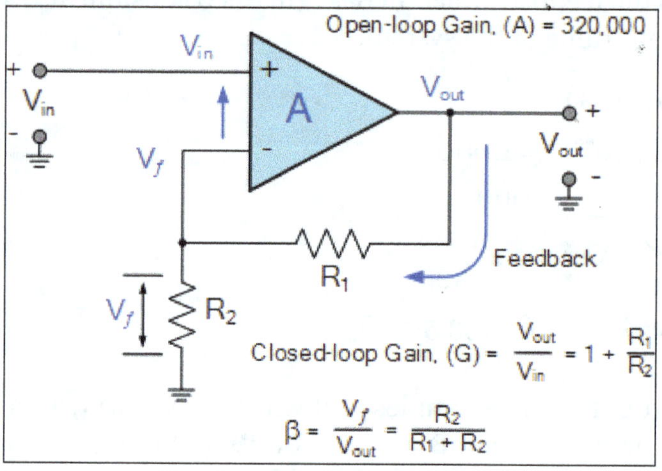

Non-inverting Op-amp Circuit.

Example: An operational amplifier with an open-loop voltage gain, A_{VOL} of 320,000 without feedback is to be used as a non-inverting amplifier. Calculate the values of the feedback resistances, R_1 and R_2 required to stabilise the circuit with a closed loop gain of 20.

The generalised closed-loop feedback equation we derived above is given as:

$$G = \frac{A}{1+\beta A}$$

By rearranging the feedback formula we get a feedback fraction, β of:

$$G(1+\beta A) = A$$

$$1+\beta A = \frac{A}{G}$$

$$\beta A = \frac{A}{G} - 1$$

$$\therefore \beta = \frac{1}{G} - \frac{1}{A}$$

Then putting the values of: A = 320,000 and G = 20, into the above equation we get the value of β as:

$$\beta = \frac{1}{G} - \frac{1}{A}$$

$$\beta = \frac{1}{20} - \frac{1}{320{,}000} = 0.05$$

Because in this case the open-loop gain of the op-amp is very high (A = 320,000), the feedback fraction, β will be roughly equal to the reciprocal of the closed-loop gain 1/G only as the value of 1/A will be incredibly small. Then β (the feedback fraction) is equal to 1/20 = 0.05.

As the resistors, R_1 and R_2 form a simple series-voltage potential divider network across the non-inverting amplifier, the closed-loop voltage gain of the circuit will be determined by the ratios of these resistances as:

$$\beta = \frac{V_f}{V_{OUT}} = \frac{R_2}{R_1 + R_2}$$

If we assume resistor R_2 has a value of 1,000Ω, or 1kΩ, then the value of resistor R_1 will be:

$$\beta = \frac{R_2}{R_1 + R_2}$$

$$\therefore R_1 = \frac{R_2 - \beta R_2}{\beta}$$

$$= \frac{1000 - (0.05 \times 1000)}{0.05} = 19{,}000\Omega = 19k\Omega$$

Then for the non-inverting amplifier circuit about to have a closed-loop gain of 20, the values of the negative feedback resistors required will be in this case, R_1 = 19kΩ and R_2 = 1kΩ, giving us a non-inverting amplifier circuit of:

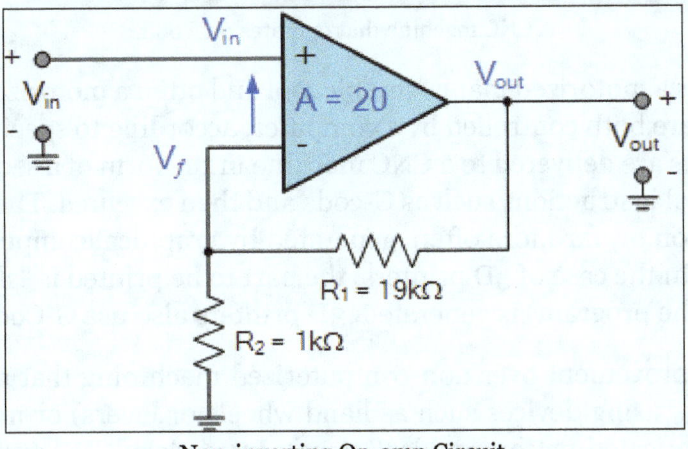

Non-inverting Op-amp Circuit.

There are many advantages to using feedback within a systems design, but the main advantages of using Negative Feedback in amplifier circuits is to greatly improve their stability, better tolerance to component variations, stabilisation against DC drift as well as increasing the amplifiers bandwidth.

Examples of negative feedback in common amplifier circuits include the resistor R_f in op-amp circuits as we have seen above, resistor, R_S in FET based amplifiers and resistor, R_E in bipolar transistor (BJT) amplifiers.

Numerical Control

Numerical control (also computer numerical control, and commonly called CNC) is the automated control of machining tools (drills, boring tools, lathes) and 3D printers by means of a computer. A CNC machine processes a piece of material (metal, plastic, wood, ceramic, or composite) to meet specifications by following a coded programmed instruction and without a manual operator.

A CNC machine that operates on wood.

A CNC machine is a motorized maneuverable tool and often a motorized maneuverable platform, which are both controlled by a computer, according to specific input instructions. Instructions are delivered to a CNC machine in the form of a sequential program of machine control instructions such as G-code and then executed. The program can be written by a person or, far more often, generated by graphical computer-aided design (CAD) software. In the case of 3D printers, the part to be printed is "sliced", before the instructions (or the program) is generated. 3D printers also use G-Code.

CNC is a vast improvement over non-computerized machining that must be manually controlled (e.g., using devices such as hand wheels or levers) or mechanically controlled by pre-fabricated pattern guides (cams). In modern CNC systems, the design of a mechanical part and its manufacturing program is highly automated. The part's mechanical dimensions are defined using CAD software and then translated into manufacturing directives by computer-aided manufacturing (CAM) software. The resulting directives are transformed (by "post processor" software) into the specific commands necessary for a particular machine to produce the component, and then are loaded into the CNC machine.

Since any particular component might require the use of a number of different tools – drills, saws, etc. – modern machines often combine multiple tools into a single "cell". In other installations, a number of different machines are used with an external controller and human or robotic operators that move the component from machine to machine. In either case, the series of steps needed to produce any part is highly automated and produces a part that closely matches the original CAD.

The first NC machines were built in the 1940s and 1950s, based on existing tools that were modified with motors that moved the tool or part to follow points fed into the system on punched tape. These early servomechanisms were rapidly augmented with analog and digital computers, creating the modern CNC machine tools that have revolutionized machining processes.

Motion is controlling multiple axes, normally at least two (X and Y), and a tool spindle that moves in the Z (depth). The position of the tool is driven by direct-drive stepper motors or servo motors in order to provide highly accurate movements, or in older designs, motors through a series of step-down gears. Open-loop control works as long as the forces are kept small enough and speeds are not too great. On commercial metalworking machines, closed loop controls are standard and required in order to provide the accuracy, speed, and repeatability demanded.

As the controller hardware evolved, the mills themselves also evolved. One change has been to enclose the entire mechanism in a large box as a safety measure, often with additional safety interlocks to ensure the operator is far enough from the working piece for safe operation. Most new CNC systems built today are 100% electronically controlled.

CNC-like systems are used for any process that can be described as movements and operations. These include laser cutting, welding, friction stir welding, ultrasonic welding, flame and plasma cutting, bending, spinning, hole-punching, pinning, gluing, fabric cutting, sewing, tape and fiber placement, routing, picking and placing, and sawing.

Examples of CNC Machines

- Mill: Translates programs consisting of specific numbers and letters to move the spindle (or workpiece) to various locations and depths. Many use G-code. Functions include: face milling, shoulder milling, tapping, drilling and some even offer turning. Today, CNC mills can have 3 to 6 axes. Most CNC mills require placing the workpiece on or in them and must be at least as big as the workpiece, but new 3-axis machines are being produced that are much smaller.

- Lathe: Cuts workpieces while they are rotated. Makes fast, precision cuts, generally using indexable tools and drills. Effective for complicated programs designed to make parts that would be infeasible to make on manual lathes. Similar control specifications to CNC mills and can often read G-code. Generally have two axes (X and Z), but newer models have more axes, allowing for more advanced jobs to be machined.
- Plasma cutter: Involves cutting a material using a plasma torch. Commonly used to cut steel and other metals, but can be used on a variety of materials. In this process, gas (such as compressed air) is blown at high speed out of a nozzle; at the same time, an electrical arc is formed through that gas from the nozzle to the surface being cut, turning some of that gas to plasma. The plasma is sufficiently hot to melt the material being cut and moves sufficiently fast to blow molten metal away from the cut.
- Electric discharge machining: (EDM), also known as spark machining, spark eroding, burning, die sinking, or wire erosion, is a manufacturing process in which a desired shape is obtained using electrical discharges (sparks). Material is removed from the workpiece by a series of rapidly recurring current discharges between two electrodes, separated by a dielectric fluid and subject to an electric voltage. One of the electrodes is called the tool electrode, or simply the “tool” or “electrode,” while the other is called the workpiece electrode, or “workpiece”.

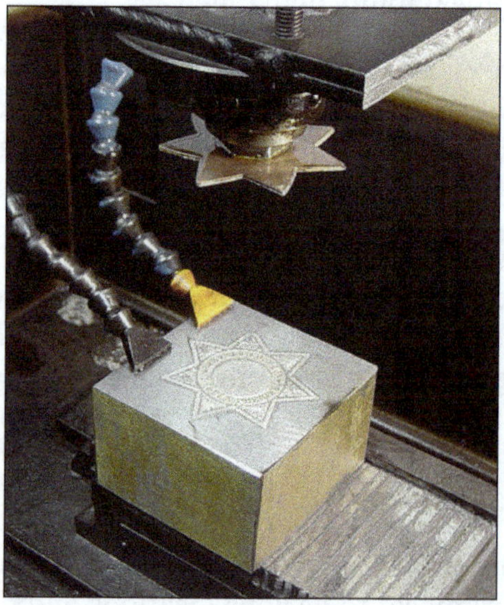

- Master at top, badge die workpiece at bottom, oil jets at left (oil has been drained). Initial flat stamping will be “dapped” to give a curved surface.

- Multi-spindle machine: Type of screw machine used in mass production. Considered to be highly efficient by increasing productivity through automation. Can efficiently cut materials into small pieces while simultaneously utilizing a diversified set of tooling. Multi-spindle machines have multiple spindles on a drum that rotates on a horizontal or vertical axis. The drum contains a drill head which consists of a number of spindles that are mounted on ball bearings and driven by gears. There are two types of attachments for these drill heads, fixed or adjustable, depending on whether the centre distance of the drilling spindle needs to be varied.

- Wire EDM: Also known as wire cutting EDM, wire burning EDM, or traveling wire EDM, this process uses spark erosion to machine or remove material from any electrically conductive material, using a traveling wire electrode. The wire electrode usually consists of brass- or zinc-coated brass material. Wire EDM allows for near 90-degree corners and applies very little pressure on the material. Since the wire is eroded in this process, a wire EDM machine feeds fresh wire from a spool while chopping up the used wire and leaving it in a bin for recycling.

- Sinker EDM: Also called cavity type EDM or volume EDM, a sinker EDM consists of an electrode and workpiece submerged in oil or another dielectric fluid. The electrode and workpiece are connected to a suitable power supply, which generates an electrical potential between the two parts. As the electrode approaches the workpiece, dielectric breakdown occurs in the fluid forming a plasma channel and small spark jumps. Production dies and moulds are often made with sinker EDM. Some materials, such as soft ferrite materials and epoxy-rich bonded magnetic materials are not compatible with sinker EDM as they are not electrically conductive.

Waterjet cutting machine for all materials.

- Water jet cutter: Also known as a "waterjet", is a tool capable of slicing into metal or other materials (such as granite) by using a jet of water at high velocity and pressure, or a mixture of water and an abrasive substance, such as sand.

It is often used during fabrication or manufacture of parts for machinery and other devices. Waterjet is the preferred method when the materials being cut are sensitive to the high temperatures generated by other methods. It has found applications in a diverse number of industries from mining to aerospace where it is used for operations such as cutting, shaping, carving, and reaming.

Other CNC Tools

Many other tools have CNC variants, including:

- Drills,
- EDMs,
- Embroidery machines,
- Lathes,
- Milling machine,
- Canned cycle,
- Wood routers,
- Sheet metal works (Turret punch),
- Tube, pipe and wire bending machines,
- Hot-wire foam cutters,
- Plasma cutters,
- Water jet cutters,
- Laser cutting,
- Oxy-fuel,
- Surface grinder,
- Cylindrical grinders,
- 3D printing,
- Induction hardening machines,
- Submerged arc welding,
- Glass cutting,
- CNC router.

Tool/Machine Crashing

In CNC, a "crash" occurs when the machine moves in such a way that is harmful to the machine, tools, or parts being machined, sometimes resulting in bending or breakage of cutting tools, accessory clamps, vises, and fixtures, or causing damage to the machine itself by bending guide rails, breaking drive screws, or causing structural components to crack or deform under strain. A mild crash may not damage the machine or tools, but may damage the part being machined so that it must be scrapped.

Many CNC tools have no inherent sense of the absolute position of the table or tools when turned on. They must be manually "homed" or "zeroed" to have any reference to work from, and these limits are just for figuring out the location of the part to work with it, and are not really any sort of hard motion limit on the mechanism. It is often possible to drive the machine outside the physical bounds of its drive mechanism, resulting in a collision with itself or damage to the drive mechanism. Many machines implement control parameters limiting axis motion past a certain limit in addition to physical limit switches. However, these parameters can often be changed by the operator.

Many CNC tools also do not know anything about their working environment. Machines may have load sensing systems on spindle and axis drives, but some do not. They blindly follow the machining code provided and it is up to an operator to detect if a crash is either occurring or about to occur, and for the operator to manually abort the active process. Machines equipped with load sensors can stop axis or spindle movement in response to an overload condition, but this does not prevent a crash from occurring. It may only limit the damage resulting from the crash. Some crashes may not ever overload any axis or spindle drives.

If the drive system is weaker than the machine structural integrity, then the drive system simply pushes against the obstruction and the drive motors "slip in place". The machine tool may not detect the collision or the slipping, so for example the tool should now be at 210mm on the X axis, but is, in fact, at 32mm where it hit the obstruction and kept slipping. All of the next tool motions will be off by −178mm on the X axis, and all future motions are now invalid, which may result in further collisions with clamps, vises, or the machine itself. This is common in open loop stepper systems, but is not possible in closed loop systems unless mechanical slippage between the motor and drive mechanism has occurred. Instead, in a closed loop system, the machine will continue to attempt to move against the load until either the drive motor goes into an overload condition or a servo motor fails to get to the desired position.

Collision detection and avoidance is possible, through the use of absolute position sensors (optical encoder strips or disks) to verify that motion occurred, or torque sensors or power-draw sensors on the drive system to detect abnormal strain when the machine should just be moving and not cutting, but these are not a common component of most hobby CNC tools.

Instead, most hobby CNC tools simply rely on the assumed accuracy of stepper motors that rotate a specific number of degrees in response to magnetic field changes. It is often assumed the stepper is perfectly accurate and never missteps, so tool position monitoring simply involves counting the number of pulses sent to the stepper over time. An alternate means of stepper position monitoring is usually not available, so crash or slip detection is not possible.

Commercial CNC metalworking machines use closed loop feedback controls for axis movement. In a closed loop system, the controller monitors the actual position of each axis with an absolute or incremental encoder. With proper control programming, this will reduce the possibility of a crash, but it is still up to the operator and programmer to ensure that the machine is operated in a safe manner. However, during the 2000s and 2010s, the software for machining simulation has been maturing rapidly, and it is no longer uncommon for the entire machine tool envelope (including all axes, spindles, chucks, turrets, tool holders, tailstocks, fixtures, clamps, and stock) to be modeled accurately with 3D solid models, which allows the simulation software to predict fairly accurately whether a cycle will involve a crash. Although such simulation is not new, its accuracy and market penetration are changing considerably because of computing advancements.

Numerical Precision and Equipment Backlash

Within the numerical systems of CNC programming it is possible for the code generator to assume that the controlled mechanism is always perfectly accurate, or that precision tolerances are identical for all cutting or movement directions. This is not always a true condition of CNC tools. CNC tools with a large amount of mechanical backlash can still be highly precise if the drive or cutting mechanism is only driven so as to apply cutting force from one direction, and all driving systems are pressed tightly together in that one cutting direction. However a CNC device with high backlash and a dull cutting tool can lead to cutter chatter and possible workpiece gouging. Backlash also affects precision of some operations involving axis movement reversals during cutting, such as the milling of a circle, where axis motion is sinusoidal. However, this can be compensated for if the amount of backlash is precisely known by linear encoders or manual measurement.

The high backlash mechanism itself is not necessarily relied on to be repeatedly precise for the cutting process, but some other reference object or precision surface may be used to zero the mechanism, by tightly applying pressure against the reference and setting that as the zero reference for all following CNC-encoded motions. This is similar to the manual machine tool method of clamping a micrometer onto a reference beam and adjusting the Vernier dial to zero using that object as the reference.

Positioning Control System

In numerical control systems, the position of the tool is defined by a set of instructions called the part program.

Positioning control is handled by means of either an open loop or a closed loop system. In an open loop system, communication takes place in one direction only: from the controller to the motor. In a closed loop system, feedback is provided to the controller so that it can correct for errors in position, velocity, and acceleration, which can arise due to variations in load or temperature. Open loop systems are generally cheaper but less accurate. Stepper motors can be used in both types of systems, while servo motors can only be used in closed systems.

Cartesian Coordinates

The G & M code positions are all based on a three dimensional Cartesian coordinate system. This system is a typical plane often seen in mathematics when graphing. This system is required to map out the machine tool paths and any other kind of actions that need to happen in a specific coordinate. Absolute coordinates are what is generally used more commonly for machines and represent the (0,0,0) point on the plane. This point is set on the stock material in order to give a starting point or "home position" before starting the actual machining.

References

- Control-engineering-historical-review-and-types-of-control-engineering; electrical4u.com, Retrieved 10 March, 2020
- Control-system, definition: whatis.techtarget.com, Retrieved 17 January, 2020
- Open-loop-system, systems: electronics-tutorials.ws, Retrieved 12 April, 2020
- Closed-loop-system: electronicshub.org, Retrieved 24 May, 2020
- What-is-pid-controller-how-it-works: electricaltechnology.org, Retrieved 08 July, 2020
- Khalil, H. K. (2002). Nonlinear Systems (3rd ed.). Upper Saddle River: Prentice Hall. ISBN 978-0-13-067389-3

Permissions

All chapters in this book are published with permission under the Creative Commons Attribution Share Alike License or equivalent. Every chapter published in this book has been scrutinized by our experts. Their significance has been extensively debated. The topics covered herein carry significant information for a comprehensive understanding. They may even be implemented as practical applications or may be referred to as a beginning point for further studies.

We would like to thank the editorial team for lending their expertise to make the book truly unique. They have played a crucial role in the development of this book. Without their invaluable contributions this book wouldn't have been possible. They have made vital efforts to compile up to date information on the varied aspects of this subject to make this book a valuable addition to the collection of many professionals and students.

This book was conceptualized with the vision of imparting up-to-date and integrated information in this field. To ensure the same, a matchless editorial board was set up. Every individual on the board went through rigorous rounds of assessment to prove their worth. After which they invested a large part of their time researching and compiling the most relevant data for our readers.

The editorial board has been involved in producing this book since its inception. They have spent rigorous hours researching and exploring the diverse topics which have resulted in the successful publishing of this book. They have passed on their knowledge of decades through this book. To expedite this challenging task, the publisher supported the team at every step. A small team of assistant editors was also appointed to further simplify the editing procedure and attain best results for the readers.

Apart from the editorial board, the designing team has also invested a significant amount of their time in understanding the subject and creating the most relevant covers. They scrutinized every image to scout for the most suitable representation of the subject and create an appropriate cover for the book.

The publishing team has been an ardent support to the editorial, designing and production team. Their endless efforts to recruit the best for this project, has resulted in the accomplishment of this book. They are a veteran in the field of academics and their pool of knowledge is as vast as their experience in printing. Their expertise and guidance has proved useful at every step. Their uncompromising quality standards have made this book an exceptional effort. Their encouragement from time to time has been an inspiration for everyone.

The publisher and the editorial board hope that this book will prove to be a valuable piece of knowledge for students, practitioners and scholars across the globe.

Index

www.ingramcontent.com/pod-product-compliance
Lightning Source LLC
LaVergne TN
LVHW080613210926
848485LV00003B/155